Information Theory and Its Application in Machine Condition Monitoring

Information Theory and Its Application in Machine Condition Monitoring

Editors

Yongbo Li
Fengshou Gu
Xihui (Larry) Liang

MDPI • Basel • Beijing • Wuhan • Barcelona • Belgrade • Manchester • Tokyo • Cluj • Tianjin

Editors
Yongbo Li
Northwestern Polytechnical
University
China

Fengshou Gu
University of Huddersfield
UK

Xihui (Larry) Liang
University of Manitoba
Canada

Editorial Office
MDPI
St. Alban-Anlage 66
4052 Basel, Switzerland

This is a reprint of articles from the Special Issue published online in the open access journal *Entropy* (ISSN 1099-4300) (available at: https://www.mdpi.com/journal/entropy/special_issues/Mach_Cond_Monit).

For citation purposes, cite each article independently as indicated on the article page online and as indicated below:

LastName, A.A.; LastName, B.B.; LastName, C.C. Article Title. *Journal Name* **Year**, *Volume Number*, Page Range.

ISBN 978-3-0365-3208-0 (Hbk)
ISBN 978-3-0365-3209-7 (PDF)

© 2022 by the authors. Articles in this book are Open Access and distributed under the Creative Commons Attribution (CC BY) license, which allows users to download, copy and build upon published articles, as long as the author and publisher are properly credited, which ensures maximum dissemination and a wider impact of our publications.

The book as a whole is distributed by MDPI under the terms and conditions of the Creative Commons license CC BY-NC-ND.

Contents

Yongbo Li, Fengshou Gu and Xihui Liang
Information Theory and Its Application in Machine Condition Monitoring
Reprinted from: *Entropy* 2022, 24, 206, doi:10.3390/e24020206 1

Gang Mao, Zhongzheng Zhang, Bin Qiao and Yongbo Li
Fusion Domain-Adaptation CNN Driven by Images and Vibration Signals for Fault Diagnosis of Gearbox Cross-Working Conditions
Reprinted from: *Entropy* 2022, 24, 119, doi:10.3390/e24010119 5

Tangbo Bai, Jialin Gao, Jianwei Yang and Dechen Yao
A Study on Railway Surface Defects Detection Based on Machine Vision
Reprinted from: *Entropy* 2021, 23, 1437, doi:10.3390/e23111437 19

Yuqing Li, Mingjia Lei, Pengpeng Liu, Rixin Wang and Minqiang Xu
A Novel Framework for Anomaly Detection for Satellite Momentum Wheel Based on Optimized SVM and Huffman-Multi-Scale Entropy
Reprinted from: *Entropy* 2021, 23, 1062, doi:10.3390/e23081062 33

Zhenhao Yan, Guifang Liu, Jinrui Wang, Huaiqian Bao, Zongzhen Zhang, Xiao Zhang and Baokun Han
A New Universal Domain Adaptive Method for Diagnosing Unknown Bearing Faults
Reprinted from: *Entropy* 2021, 23, 1052, doi:10.3390/e23081052 61

Yancai Xiao, Jinyu Xue, Mengdi Li and Wei Yang
Low-Pass Filtering Empirical Wavelet Transform Machine Learning Based Fault Diagnosis for Combined Fault of Wind Turbines
Reprinted from: *Entropy* 2021, 23, 975, doi:10.3390/e23080975 75

Zhongshuo Hu, Jianwei Yang, Dechen Yao, Jinhai Wang and Yongliang Bai
Subway Gearbox Fault Diagnosis Algorithm Based on Adaptive Spline Impact Suppression
Reprinted from: *Entropy* 2021, 23, 660, doi:10.3390/e23060660 91

Sixiang Jia, Jinrui Wang, Xiao Zhang and Baokun Han
A Weighted Subdomain Adaptation Network for Partial Transfer Fault Diagnosis of Rotating Machinery
Reprinted from: *Entropy* 2021, 23, 424, doi:10.3390/e23040424 109

Juhui Wei, Zhangming He, Jiongqi Wang, Dayi Wang and Xuanying Zhou
Fault Detection Based on Multi-Dimensional KDE and Jensen–Shannon Divergence
Reprinted from: *Entropy* 2021, 23, 266, doi:10.3390/e23030266 123

Yancai Xiao, Jinyu Xue, Long Zhang, Yujia Wang and Mengdi Li
Misalignment Fault Diagnosis for Wind Turbines Based on Information Fusion
Reprinted from: *Entropy* 2021, 23, 243, doi:10.3390/e23020243 147

Wentao Mao, Bin Sun, Liyun Wang and Naiqin Feng
A New Deep Dual Temporal Domain Adaptation Method for Online Detection of Bearings Early Fault
Reprinted from: *Entropy* 2021, 23, 162, doi:10.3390/e23020162 167

Editorial

Information Theory and Its Application in Machine Condition Monitoring

Yongbo Li [1,*], Fengshou Gu [2] and Xihui Liang [3]

1 School of Aeronautics, Northwestern Polytechnical University, Xi'an 710072, China
2 Centre for Efficiency and Performance Engineering (CEPE), University of Huddersfield, Queensgate, Huddersfield HD1 3DH, UK; F.Gu@hud.ac.uk
3 Department of Mechanical Engineering, University of Manitoba, Winnipeg, MB R3T 5V6, Canada; xihui.liang@umanitoba.ca
* Correspondence: yongbo@nwpu.edu.cn

Introduction

Rotating machinery is part and parcel of modern industrial applications. Due to their continuous nature of operation in harsh and varying operating conditions, rotating machinery is much more prone to failure. Failure to diagnose the fault on time can lead to catastrophic effects from both a financial and safety point of view. In this context, research on entropy theory and its application in machine condition monitoring is very interesting, as it can bring some new insights into the prognostic and health management of academic and industrial files. Information theory, as one of the greatest discoveries in the history of science, has formed more than ten different calculation forms after more than 70 years of development. Information theory has been widely used in the feature extraction field due to its merits of independence with prior knowledge, the lack of a need to preprocess, and ease of performance. It has powerful ability for feature representation, has received extensive attention from both academic and industrial attentions.

This Special Issue in Entropy aims to collect recent research results and the latest developments in condition monitoring techniques by means of information theory. This collection contains 10 papers that represent the state of the art application of information theory in the field of condition monitoring.

To resolve the dilemma of diagnosing non-structural failures, Mao et al. proposed a fusion domain-adaptation convolutional neural network (FDACNN) to combine infrared thermal images and vibration signals for the diagnosis of both structural and non-structural failures under various working conditions [1]. They designed an adversarial network to recognize of the health condition of structural and non-structural faults in the unlabeled target domain. The results suggest that the proposed FDACNN method performs best in the cross-domain fault diagnosis of gearboxes via multi-source heterogeneous data.

Bai et al. proposed a new method based on an improved YOLOv4 for railway surface defect detection [2]. In this method, MobileNetv3 was used as the backbone network of YOLOv4 to extract image features, and deep separable convolution as applied on the PANet layer in YOLOv4 to realize the lightweight network and achieve real-time detection of the railway surface.

Xiao et al. diagnosed the different types of misalignment fault, such as parallel misalignment, angular misalignment and integrated misalignment, of a wind turbine by means of information fusion [3]. The concept of information fusion was realized by fusing different types of sensor data, such as vibration, temperature and current. Finally, fault diagnosis of the wind turbine was achieved from the fused information with the help of Dempster–Shafer evidence theory.

Jia et al. showed a partial transfer fault diagnosis model based on a weighted subdomain adaptation network (WSAN) [4]. This model paid more attention to the important local data distribution, while aligning the global distribution through a specially designed weighted local maximum mean discrepancy (WLMMD), which was able to align relevant distributions of domain-specific layer activations across different domains.

Targeting the challenges regarding weak fault signal, coupling among different dimensions of the collected signal, and scarcity of fault datasets, Wei et al. proposed a novel fault detection based on multi-dimensional KDE and Jensen–Shannon divergence [5]. Addressing the limitations of the conventional KDE method regarding information loss for multidimensional problems, in this research, it was extended to a multidimensional version for tackling the weak fault signal and coupling problem of the collected signal. Later, Jensen–Shannon divergence was used for unknown fault detection.

Hu et al. proposed a subway gearbox fault diagnosis algorithm based on adaptive spline impact suppression [6]. Signals collected from the gearbox in a subway are often associated with the impacts between wheelsets and rail joint gaps. A long time-series is more often affected by unsteadiness due to the wheel-rail impacts. Motivated by this phenomenon, in this paper, long times-series were segmented adaptively into short time series, in order to suppress the effect of the high amplitude of the shock response signal. The adaptive segmentation of the original signal was achieved using a cubic spline interpolation algorithm. The proposed method was verified by data collected from a real Beijing subway line.

Yan et al. proposed a universal domain adaptation method to recognize unknown fault types in the target domain [7]. They took into account the discrepancy of the fault features shown by different fault types, and formed the feature center for fault diagnosis by extracting the features of the samples of each fault type.

Wind turbines are one of the major sources of green energy. However, the majority of the wind turbine fault diagnosis research is focused on single faults rather than combined faults. In order to fill this gap, in this Special Issue, Xiao et al. proposed a machine-learning-based approach with the help of the feature extracted by a novel low-pass filtering empirical wavelet transform [8].

Li et al. proposed a novel anomaly detection framework for a satellite momentum wheel [9]. Aimed at the lack of research on simulation data, and the scarcity of research on real telemetry data, the proposed framework was able to detect anomalies in a satellite momentum wheel, based on the features extracted by a newly proposed Huffman-multi-scale entropy method. The extracted features were then classified by an optimized support vector machine algorithm.

Mao et al. suggested a new online detection method, called the deep dual-temporal domain adaptation model (DTDA), to effectively extract a domain-invariant temporal feature representation for online early-fault detection of bearings [10]. They developed a new state assessment method to determine the period of the normal state and that of the degradation state for whole-life degradation sequences. A health indicator of target bearings was also built based on the DTDA features, to intuitively evaluate the detection results.

The editors extend their heartiest gratitude toward all of the the authors for their excellent contribution to this Special Issue. Special thanks to all of the anonymous reviewers for their time and feedback provided to the authors. Additionally, we sincerely thank the publishers, editors and all of the members of the Entropy editorial board for facilitating this opportunity to present all of the works.

Funding: This research received no external funding.

Acknowledgments: We express our thanks to the authors of the above contributions, and to the journal *Entropy* and MDPI for their constant and precious support during this work.

Conflicts of Interest: The authors declare no conflict of interest.

References

1. Mao, G.; Zhang, Z.; Qiao, B.; Li, Y. Fusion domain-adaptation CNN driven by images and vibration signals for fault diagnosis of gearbox cross-working conditions. *Entropy* **2022**, *24*, 119. [CrossRef] [PubMed]
2. Bai, T.; Gao, J.; Yang, J.; Yao, D. A study on railway surface defects detection based on machine vision. *Entropy* **2021**, *23*, 1437. [CrossRef] [PubMed]
3. Xiao, Y.; Xue, J.; Zhang, L.; Wang, Y.; Li, M. Misalignment fault diagnosis for wind turbines based on information fusion. *Entropy* **2021**, *23*, 243. [CrossRef] [PubMed]
4. Jia, S.; Wang, J.; Zhang, X.; Han, B. A weighted subdomain adaptation network for partial transfer fault diagnosis of rotating machinery. *Entropy* **2021**, *23*, 424. [CrossRef] [PubMed]
5. Wei, J.; He, Z.; Wang, J.; Wang, D.; Zhou, X. Fault detection based on multi-dimensional KDE and Jensen–Shannon divergence. *Entropy* **2021**, *23*, 266. [CrossRef] [PubMed]
6. Hu, Z.; Yang, J.; Yao, D.; Wang, J.; Bai, Y. Subway gearbox fault diagnosis algorithm based on adaptive spline impact suppression. *Entropy* **2021**, *23*, 660. [CrossRef] [PubMed]
7. Yan, Z.; Liu, G.; Wang, J.; Bao, H.; Zhang, Z.; Zhang, X.; Han, B. A new universal domain adaptive method for diagnosing unknown bearing faults. *Entropy* **2021**, *23*, 1052. [CrossRef] [PubMed]
8. Xiao, Y.; Xue, J.; Li, M.; Yang, W. Low-pass filtering empirical wavelet transform machine learning based fault diagnosis for combined fault of wind turbines. *Entropy* **2021**, *23*, 975. [CrossRef] [PubMed]
9. Li, Y.; Lei, M.; Liu, P.; Wang, R.; Xu, M. A novel framework for anomaly detection for satellite momentum wheel based on optimized SVM and Huffman-Multi-Scale entropy. *Entropy* **2021**, *23*, 1062. [CrossRef] [PubMed]
10. Mao, W.; Sun, B.; Wang, L. A new deep dual temporal domain adaptation method for online detection of bearings early fault. *Entropy* **2021**, *23*, 162. [CrossRef] [PubMed]

Article

Fusion Domain-Adaptation CNN Driven by Images and Vibration Signals for Fault Diagnosis of Gearbox Cross-Working Conditions

Gang Mao, Zhongzheng Zhang, Bin Qiao and Yongbo Li *

MIIT Key Laboratory of Dynamics and Control of Complex System, School of Aeronautics, Northwestern Polytechnical University, Xi'an 710072, China; mg0207@yeah.net (G.M.); zhangzhongzheng@mail.nwpu.edu.cn (Z.Z.); qiaobin@mail.nwpu.edu.cn (B.Q.)
* Correspondence: yongbo@nwpu.edu.cn

Abstract: The vibration signal of gearboxes contains abundant fault information, which can be used for condition monitoring. However, vibration signal is ineffective for some non-structural failures. In order to resolve this dilemma, infrared thermal images are introduced to combine with vibration signals via fusion domain-adaptation convolutional neural network (FDACNN), which can diagnose both structural and non-structural failures under various working conditions. First, the measured raw signals are converted into frequency and squared envelope spectrum to characterize the health states of the gearbox. Second, the sequences of the frequency and squared envelope spectrum are arranged into two-dimensional format, which are combined with infrared thermal images to form fusion data. Finally, the adversarial network is introduced to realize the state recognition of structural and non-structural faults in the unlabeled target domain. An experiment of gearbox test rigs was used for effectiveness validation by measuring both vibration and infrared thermal images. The results suggest that the proposed FDACNN method performs best in cross-domain fault diagnosis of gearboxes via multi-source heterogeneous data compared with the other four methods.

Keywords: deep learning; fault diagnosis; multi-source heterogeneous fusion; gearbox; transfer learning

1. Introduction

The gearboxes play an irreplaceable role in the mechanical power system, which usually works in harsh and complex environments [1,2]. The failure of gearboxes may cause unexpected accidents and economic losses. Therefore, accurately identifying and diagnosing faults is the key to ensuring that gearboxes normally operate [3].

The intelligent diagnosis method has received wide attention from researchers because of its ability to detect faults automatically and is not limited by manual experience. The methods based on deep learning are particularly prominent because they can adaptively learn the fault information hidden in the collected signals, such as long short-term memory network (LSTM) [4], recurrent neural network (RNN) [5] and convolutional neural network (CNN) [6]. In addition, some extended models based on standard deep learning models are proposed for rotating machinery fault diagnosis, such as deep convolutional auto-encoder (DCAE) [7], CNN with capsule network [8] and multiscale CNN [9], etc. Yao et al. [10] proposed a stacked inverted residual CNN (SIRCNN), which had stable and reliable fault diagnosis accuracy. Shao et al. [11] established an ensemble deep auto-encoder (EDAE), which consists of several DAEs with different activation functions. The results indicated that it has good accuracy in rolling bearing fault diagnosis. However, the methods mentioned above assume that adequate high-quality data collected from the concerned machine are available for estimating underlying data distributions. In addition, these methods need training and testing data drawn from the same probability distribution [12]. In actual applications, it is impractical to obtain a large amount of labeled data.

In addition, the performance of the aforementioned methods may decrease in recognizing unlabeled data collected from another machine or different working conditions due to data discrepancy [13].

Transfer learning can transfer learned knowledge to related machinery or fields, and it is widely applied to address the above-mentioned cross-domain fault diagnosis problem. Currently, parameter transfer and feature domain-adaptation are two popular transfer learning implementation methods. Parameter transfer is suitable for scenarios where only a few labeled data from the target domain are available, but it is not enough to train the model. Qian et al. [14] proposed a method for rolling bearing fault diagnosis under variable working conditions by transferring the parameters of the stack auto-encoder (SAE). Chen et al. [15] proposed the transfer neural network to diagnose the faults of the rotary machinery, which pre-trains a 1D-CNN with the source data and then uses the limited target data to fine-tune the model to obtain a transfer convolutional neural network. However, in most practical applications, there is no available labeled target data to participate in the model training process. Domain adaptation techniques based on feature transfer have been much preferred in this case. One implementation of domain adaption is to add a domain adaptation term to the loss function, such as Maximum Mean Discrepancy (MMD) [16–18] and Wasserstein distance [19]. Another implementation of domain adaption is through domain adversarial training, in which a feature extractor aims to extract common features from both source and target domain by adversarial training [20–22]. In addition to this, in order to further improve transfer and generalization capabilities of the models, multiple source domains of data are used to extract transferable features, which are used to diagnose the faults of rotating machinery [23–25].

However, most existing studies on transfer diagnosis mainly focus on single-channel signals with vibration signals as the mainstay. This is because the vibration signal can be collected by the acceleration sensor attached to the surface of the component, which is sensitive to the impact caused by structural damage, such as gear fracture and bearing outer race crack. For some non-structural faults, such as gear box oil shortages, vibration signals are not sensitive to them. These failures can also cause serious consequences and should not be ignored. Infrared thermal image can perfectly reflect non-structural fault information and is widely applied in fault diagnosis [26,27]. However, the single infrared thermal image is very sensitive and is easily affected by external factors such as oil temperature [28]. Therefore, the fault diagnosis method based on multi-source heterogeneous data fusion is an issue worthy of study. Bai et al. [29] proposed a method for coupling fault diagnosis of rotary machinery by using infrared images and vibration signals, in which the enhanced infrared thermal image and two-dimensional vibration signals are spliced and inputted into CNN to obtain final diagnosis result. Shao et al. [30] pre-trained multiple novel SAEs using multisensory signals from the source domain and finetuned each novel SAE using a target domain sample. The diagnosis result is obtained by a modified voting strategy. In the above research studies, multi-source heterogeneous signals are widely applied in fault diagnosis since they can supply abundant fault information. However, it is rare to use infrared thermal images and vibration signals to diagnose structured and unstructured failure states in unlabeled target domains.

This paper proposed a fusion domain-adaptation CNN (FDACNN) driven by images and vibration signals. An FDACNN consists of two main stages: data-level fusion and domain-adaptation network training. In the stages of data-level fusion, raw signals are transformed into frequency and squared envelope spectrum, and they are arranged into two-dimensional format. Two-dimensional format data are combined with the infrared thermal image to form fusion data samples for model training. In the stages of domain-adaptation network training, a features extractor, a domain discriminator and a state classifier are constructed. After a number of adversarial training, the domain invariant features can be extracted from fusion samples and used for the classification of health states. In actual industrial gearbox, both the accelerators and infrared camera can be installed to collect the infrared images and vibration signals. The vibration signal can be used to effectively diagnose structural failures such as tooth breakage, tooth missing, and gear wear. Moreover, the infrared thermal image is sensitive to non-structural failures, such as oil shortage and oil temperature exorbitant. In this study, more comprehensive features can be extracted from infrared thermal image and vibration signals than a single sensor. Moreover, the proposed method has lower calculation complexity, which can rely in the online fault diagnosis of gearbox. The main contributions and insights of this study are listed below:

(1) A data-level multi-source heterogeneous fusion scheme is proposed. The frequency and squared envelope spectrum can more clearly reflect fault information contained in the vibration signal. The fusion of the preprocessed vibration signal and the infrared thermal image makes the fault information in the training sample more abundant and obvious.

(2) A fusion domain-adaptation CNN fault diagnosis method for gearboxes is explored. It can extract domain invariant features from the fusion information of vibration signals and infrared thermal images and implement gearbox fault diagnosis in an unlabeled target domain.

The rest of this article is arranged as follows: Section 2 presents preliminary and basic knowledge. The details of the proposed FDACNN are provided in Section 3. Section 4 validates the proposed method and analyzes the results. Finally, the conclusion in Section 5 brings the study to a close.

2. Preliminaries

2.1. Squared Envelope Spectrum

Demodulation analysis methods can extract and identify fault characteristic frequencies from the resonance band. Envelope analysis is widely applied to acquire the harmonics of characteristic frequencies, such as the squared envelope spectrum (SES). Since the faulty rotating machinery vibration signals usually contain second-order cyclostationary (CS2) components, they are often used to extract fault features. CS2 is commonly calculated using SES, and the formulas can be expressed as follows:

$$SES(\alpha) = \left| \frac{1}{L} \sum_{n=0}^{L-1} |\tilde{x}(n)|^2 e^{-j2\pi n\alpha/F_s} \right|^2 = \left| DFT\left(|\tilde{x}(n)|^2 \right) \right|^2 \quad (1)$$

where α represents cyclic frequency, and F_s denotes the sample frequency. $\tilde{x}(n)$ is converted by the Hilbert transform from time-domain vibration signals. $DFT(\bullet)$ represents the discrete flourier transform, and it is formulated as follows:

$$DFT[x(n)] = \sum_{n=0}^{L-1} x(n) e^{-j \cdot k \cdot n \frac{2\pi}{L}} \quad (2)$$

where $x(n)$ represents the signal sequence $[0\ L-1]$. Thus, CS2 components is acquired with cyclic frequency α.

2.2. Convolutional Neural Network

Convolutional neural network is a typical deep feed-forward artificial neural network that can be used to process time sequences and images by convolution operation. This operation can reduce the number of weights and biases to decrease the complexity of the model. A standard CNN consists of convolution layer, pooling layer, fully connected layer and classification layer. In a convolutional layer, multiple convolutional kernels are used to convolution the input, and the weights and bias are shared between hidden neurons. The process in the convolutional layer can be expressed as follows:

$$z_n^l = f^l (\sum_k x_k^{l-1} * w_n^l + b_n^l) \quad (3)$$

where $\mathbf{x}_k^{l-1}$ is the k-th input sample in the l-1 layer. * represents convolution operation. $\mathbf{w}_n^l$ and $\mathbf{b}_n^l$ denote the weight and corresponding bias, respectively. Additionally, $f^l(\cdot)$ represents the activation function.

A pooling layer usually follows the convolutional layer, and the subsampling operations is employed to reduce the spatial dimension for reducing overfitting risk. Mathematically, a maximum pooling operation is defined as follows:

$$po_j = \max_{i \in m_j}\{\mathbf{c}_j(i)\} \quad (4)$$

where $\mathbf{c}_j$ represent the j-th location, and po_j is the output of the pooling. Moreover, average pooling and stochastic pooling are also usually used in pooling layer.

After several convolutional and pooling layers, the fully connected layer immediately converts the output matrix into a row or column. The last layer is usually served by a softmax output layer in which a softmax function is utilized to predict the probability of each target.

2.3. Deep Adversarial Convolution Neural Network

Generally, a deep adversarial convolution neural network usually consists of a feature extractor G_f, a domain discriminator G_d and a classifier G_c [13]. The feature extractor, which is a competitor in the DACNN, is typically served by several convolution blocks or fully connected layers. It can be expressed as $G_f = G_f(x, \theta_f) : x \to R^D$ with parameter θ_f, which indicates that the input sample x is transformed into D-dimensional features. In addition, the domain discriminator (binary classifier) is treated as the opponent, which is expressed as $G_d = G_d(G_f(x), \theta_d)$ with parameters θ_d. Inputting the source and target samples into the feature extractor and the output is further distinguished by the domain discriminator G_d. The binary cross entropy (BCE) loss is taken as objective function, which can be described as follows:

$$L(G_d(G_f(x_i)), d_i) = d_i \log \frac{1}{G_d(G_f(x_i))} + (1 - d_i) \times \log \frac{1}{1 - G_d(G_f(x_i))} \quad (5)$$

where d_i denotes the binary variable for x_i. By conducting adversarial training between the two parts, feature extractor G_f tends to extract common features from two types of data and makes the domain discriminator difficult to distinguish in terms of zero or one. Thus, the model can perform well on both the source and target datasets. Assuming n samples in the source domain dataset and $N-n$ samples in the target domain dataset, the objective function is expressed as follows:

$$E(\theta_f, \theta_d) = -\left(\frac{1}{n}\sum_{i=1}^n L_d^i(\theta_f, \theta_d) + \frac{1}{N-n}\sum_{i=n+1}^N L_d^i(\theta_f, \theta_d)\right) \quad (6)$$

where $L_d^i(\theta_f, \theta_d) = L_d(G_d(G_f(x_i, \theta_f), \theta_d), d_i)$, and this equation includes a maximization problem with respect to θ_d and a minimization problem with respect to θ_f.

Additionally, all the labeled samples should be supervised and trained to ensure the accuracy of the diagnosis in the adversarial procedure. Therefore, a classifier is established, and it is expressed as $G_y = G_y(G_f(x), \theta_y) : R^D \to R^L$ with parameters θ_y, in which L is the number of classes. Cross-entropy loss is applied in the Softmax function, and it can be described as follows:

$$L_y(G_y(G_f(x_i)), y_i) = \log \frac{1}{G_y(G_f(x_i))_{y_i}} \quad (7)$$

By adding the Equation (7) to objective function (6), the optimization objective can be expressed as follows:

$$E(\theta_f, \theta_y, \theta_d) = \frac{1}{n}\sum_{i=1}^{n} L_y^i(\theta_f, \theta_y) - \lambda(\frac{1}{n}\sum_{i=1}^{n} L_d^i(\theta_f, \theta_d) + \frac{1}{N-n}\sum_{i=n+1}^{N} L_d^i(\theta_f, \theta_d)) \quad (8)$$

where $L_y^i(\theta_f, \theta_y) = L_y(G_y(G_f(x_i, \theta_f), \theta_y), y_i)$. The entire training process of DANN is to optimize the parameters θ_f, θ_y and θ_d, and it is expressed as follows.

$$(\hat{\theta}_f, \hat{\theta}_y) = \underset{\theta_f, \theta_y}{\mathrm{argmax}} E(\theta_f, \theta_y, \hat{\theta}_d) \quad (9)$$

$$\hat{\theta}_d = \underset{\theta_d}{\mathrm{argmax}} E(\hat{\theta}_f, \hat{\theta}_y, \theta_d) \quad (10)$$

In the training stage, parameter updates are implemented in the opposite direction to the gradient in the adversarial process. The update in the domain discriminator G_d is to reduce the loss with the purpose of improving the discriminative ability. However, the update in the feature extractor G_f is to maximize the loss to fool the discriminator. In order to frame a flexible implement of the stochastic gradient descent (SGD) algorithms in the training of DACNN, we use a circuitous method by rewriting the loss as $L'^i_d(\theta_f, \theta_d) = L_d(G_d(G_f(x_i, \theta_f), \theta_d), 1 - d_i)$ in updating the feature extractor parameters. Maximizing $L_d^i(\theta_f, \theta_d)$ can be accomplished by minimizing $L'^i_d(\theta_f, \theta_d)$. During backpropagation, the features extractor takes the gradient of the recalculated $L'^i_d(\theta_f, \theta_d)$ from the domain discriminator and updates its parameters with SGD. Overall, the update rales of parameters θ_f, θ_y and θ_d can be formulated as follows:

$$\theta_f \leftarrow \theta_f - \mu\,(\frac{\partial L_y^i}{\partial \theta_f} + \lambda \frac{\partial L'^i_d}{\partial \theta_f}) \quad (11)$$

$$\theta_d \leftarrow \theta_d - \mu \frac{\partial L_d^i}{\partial \theta_d} \quad (12)$$

$$\theta_y \leftarrow \theta_y - \mu \frac{\partial L_y^i}{\partial \theta_y} \quad (13)$$

where μ represents the learning rate.

By the above optimization process, DACNN tends to train a feature extractor G_f that can extract suitable representations from input samples (either source domain datasets or target domain datasets), which can be classified accurately by the classifier G_y but weakens the ability of the domain discriminator G_d to differentiate a sample from the source or target domain datasets. In the phases of testing, domain insensitive features will be extracted by feature descriptor G_f and fed into the classifier G_y to identify the states immediately.

3. The Proposed Method
3.1. Data-Level Fusion

Data-level fusion is a relatively direct fusion method that can retain the effective fault information hidden in the measured signals and reduce the complexity of the model. Therefore, a data-level fusion strategy is designed in this study to fuse one-dimensional vibration signals and two-dimensional infrared thermal images.

In general, the measured time-domain vibration signals are feeble and inadequate, particularly in the early stages of a failure. Although many deep learning methods only use raw time-domain signals for fault diagnosis, they rely heavily on deep learning structure. It is considered that frequency features play a significant role in rotating machinery failure diagnostics. The frequency domain signal and the squared envelope spectrum, in particular, are widely used in classical signal processing methods [31,32]. Therefore, measured raw signals are transformed to acquire frequency domain signals by fast Fourier transform (FFT) and CS2 by squared envelope spectrum in this study. Then, they are reshaped into two-dimensional matrix as part of the input of the convolution layer as shown in Figure 1.

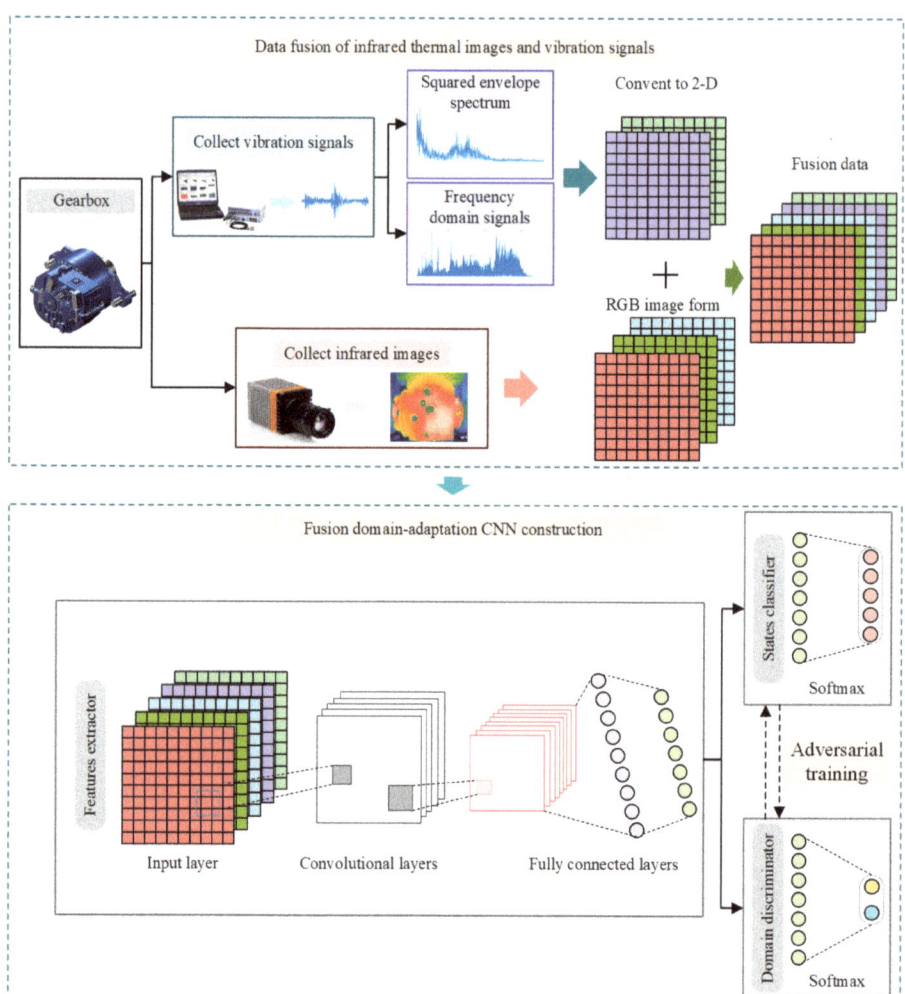

Figure 1. The procedures of the proposed method.

After that, the RGB 3-channels of each infrared thermal image are combined with the 2-dimensional frequency domain signal and the squared envelope spectrum to synthesize a 5-channel data. The 5-channel fusion data will be used for subsequent domain adaptation CNN training.

3.2. Fusion Domain-Adaptation CNN Construction

Domain adaptation based on adversarial networks is an effective approach for cross-domain fault diagnosis. In this section, fusion 5-channel data are used to train a domain-adaptation CNN for cross-domain fault diagnosis of gearboxes.

Firstly, a feature extractor is constructed, which contains multiple convolutional blocks and several fully connected layers. The feature extractor is used to extract domain-insensitive features from 5-channel fusion data. The extracted features are fed into the state classifier to recognize health states. Meanwhile, the extracted features are inputted into the domain discriminator to distinguish whether they are from the source or target domain.

In the training process, the feature extractor minimizes the state classification loss and maximizes the domain discrimination loss so that it can extract features that are not only not sensitive to the domain and make the state classifier easy to classify. As the opponent, the domain discriminator aims to minimize domain discrimination loss so that it can distinguish the feature from the source or target domain. Finally, through a large amount of adversarial training, the diagnosis model composed of feature extractor and state classifier can recognize fault states accurately with multi-source heterogeneous data under cross-working conditions.

3.3. Procedures of Proposed Fusion Domain-Adaptation CNN

This section presents the summaries of the proposed fusion domain-adaptation CNN as shown in Figure 1. The main procedures are described as follows.

Collect the infrared thermal images and raw vibration signals from the concerned gearboxes under different working conditions and divide them into labeled source domain samples and unlabeled target domain samples.

Convert the raw time-domain signals into frequency domain signals and the squared envelope spectrum and arrange them into matrixes.

Fuse the RGB 3-channels of infrared thermal image and two matrixes (frequency domain and squared envelope spectrum) to obtain 5-channel fusion samples.

Train the FDACNN model using 5-channel fusion samples by adversarial training.

Test the performance of the proposed FDACNN model by using the remaining samples from the target domain.

4. Experimental and Result Discussion

4.1. Dataset Descriptions

Test data are from a compound gear failure experiment performed on a helical gearbox called Spectra Quest Mechanical Failure Simulator (MFS) from Northwestern Polytechnical University lab [33,34]. The experiment system and the layout of the experiment rig are shown in Figure 2a,b, respectively. The experiment system mainly consists of an AC motor, two gearboxes and a generator. The infrared camera is fixed on the front of gearbox 1 to collect the infrared thermal image. The detailed parameters of the infrared camera are listed in Table 1. Vibration signals are collected by an acceleration sensor mounted on the surface of the gearbox 1. The sample frequency is 12.8 kHz, and the motor speed is 3000 rpm. In this experiment, there are two kinds of lubricating oil, i.e., EP 320 and EP 100. EP 320 lubricant viscosity is 320cSt @ 40 °C, EP 100 lubricant viscosity is 100cSt @ 40 °C. The EP 320 lubricant is applied in this article.

In this study, five different health states are introduced, including a normal state, two structural fault states (TB 50 and TB 100) and two non-structural fault states (OS 1500 and OS 2000). "TB 50" and "TB 100" refer to 50% and 100% tooth breakage in driving gear, respectively. Based on the baseline oil of 2600 mL, "OS 1500" and "OS 2000" refer to the reduction of 600 mL and 1100 mL of oil from GB1, respectively. Vibration signals and infrared images were collected under four different loads of 0%, 30%, 70% and 100% (L0, L30, L70 and L 100). For each load, vibration signals in each state were divided into 800 samples with 2048 data points. Four-hundred and eighty samples were randomly selected as tests, and the remaining 320 samples were used to train. Similarly, 480 infrared images were used to train, and the other 320 were used to test. The size of each infrared thermal image is 64 × 32. The details of the dataset are listed in Table 2.

4.2. Implementation Details

At first, data-level fusion strategy is used to fuse infrared thermal images and vibration signals. The measured raw samples in five different health states are transformed to acquire frequency domain signals by FFT and CS2 by squared envelope spectrum in this study. Under L0 load, the three kinds of signal waveforms of different health states are shown in Figure 3. As shown in Figure 3, the left column is the measured raw signals, and the middle column and the right column are the corresponding spectral distribution and squared envelope spectrum, respectively. It can be observed that the time-domain characteristics and frequency domain characteristics of each health state are relatively similar and difficult to distinguish. Then, frequency domain signals and CS2 sequence are arranged into $2 \times 64 \times 32$ formats.

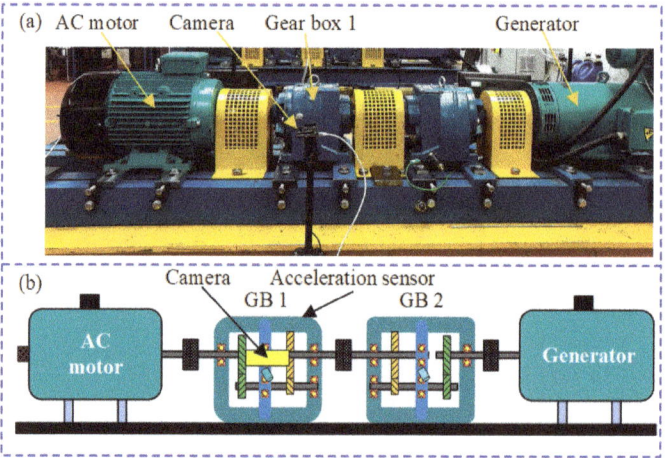

Figure 2. The gear box fault simulator system: (**a**) the experimental test rig; (**b**) the layout of the test rig.

Table 1. The detailed parameters of the infrared camera.

Parameters	Values
Alg type	PHE
Frame rate	25 fps
Temperate measurement range	−25 °C~260 °C
Environment temperature	18.9 °C
Thermal sensitivity	0.050 °C
Image resolution	384 × 288
Contrast	50
Brightness	50
Gain	2
Palette	rainbow

Table 2. 5 health states of gearbox.

Label	Health States	The Number of Training/Testing Samples
1	Normal	480/320
2	TB 50	480/320
3	TB 100	480/320
4	OS 1500	480/320
5	OS 2000	480/320

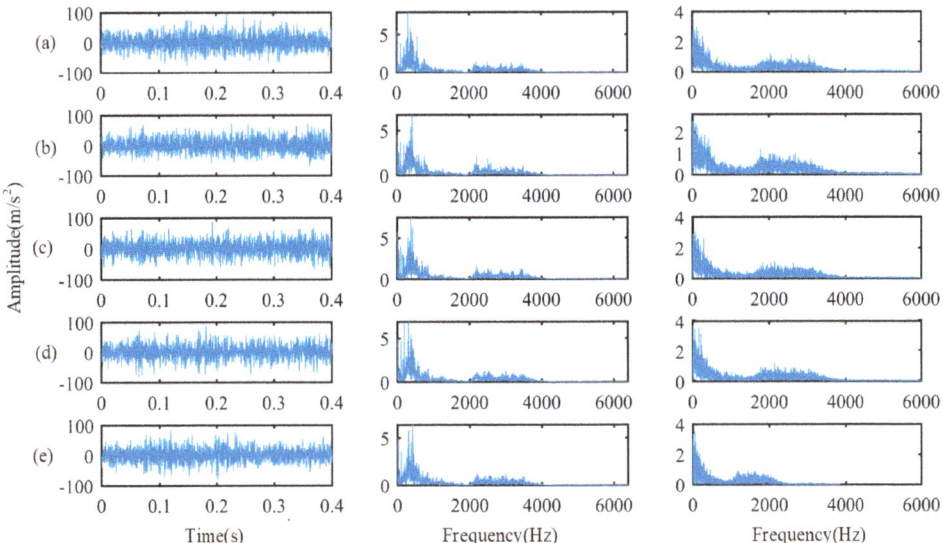

Figure 3. Raw signals, spectral distribution and squared envelope spectrum of different health states. (**a**) Normal; (**b**) TB 50; (**c**) TB 100; (**d**) OS 1500; (**e**) OS 2000.

Each infrared thermal image has RGB channels, i.e., 3 × 64 × 32 formats. The collected infrared thermal images of each health state under L0 are shown in Figure 4. From Figure 4, we can observe that the images of normal, OS 1500 and OS 2000 are relatively similar, and the images of TB 50 and TB 100 are relatively similar. However, it is still very difficult to visually distinguish concrete health states. RGB channels of the infrared thermal image (3 × 64 × 32) will be combined with the frequency domain signals and CS2 (2 × 64 × 32), and the 5-channel fusion samples (5 × 64 × 32) are obtained to train DACNN.

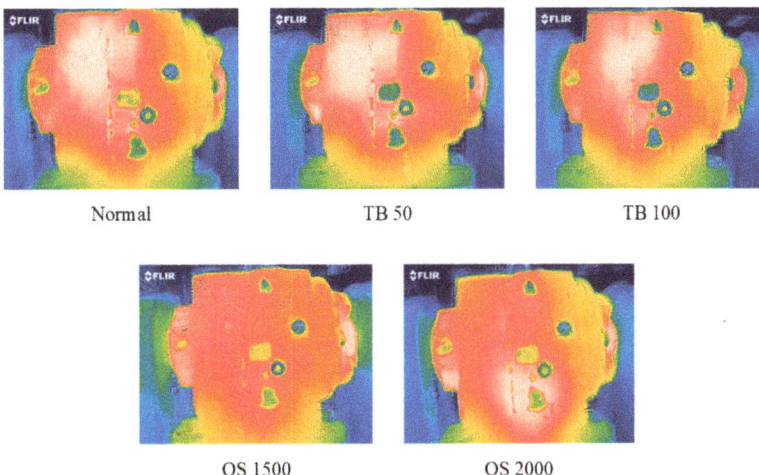

Figure 4. The infrared thermal image of different health states.

In DACNN, a feature extractor, a domain discriminator and a state classifier were constructed, and the structures of those are listed in Table 3. The DACNN is trained by adversarial training using fusion data. In order to illustrate the robustness of the proposed method, multiple test tasks are designed. The concrete setting of different tasks and the results are listed in Table 4. It can be observed that the proposed method has good performance among the five test tasks, especially the accuracy of reaching 100.00% in T1, T3 and T4. In task T5, the transfer span is larger from load L0 to L100. The accuracy of the proposed method can still reach 96.67%. This suggests that the fusion of infrared thermal images and vibration signals to implement cross-domain fault diagnosis has good performance, and the result is relatively robust.

Table 3. The structures of features extractor, domain discriminator and states classifier.

Model	Layer	Filter Number	Size of Kernel	Output Size	Stride	Padding	Active Function
	Conv2d 1	8	3 × 3	8 × 62 × 30	[1,1]	0	ReLU
	BN 1	8	-	8 × 62 × 30	-	-	-
	MaxPool2d 1	8	2 × 2	8 × 31 × 15	[2,2]	-	-
	Conv2d 2	16	3 × 3	16 × 29 × 13	[1,1]	0	ReLU
Features	BN 2	16	-	16 × 29 × 13	-	-	-
extractor	MaxPool2d 2	16	2 × 2	16 × 14 × 6	[2,2]	-	-
	FC 1	-	-	680	-	-	-
	FC 2	-	-	300	-	-	ReLU
	FC 3	-	-	56	-	-	ReLU
	FC 4	-	-	28	-	-	ReLU
States classifier	FC	-	-	5	-	-	Softmax
Domain discriminator	FC	-	-	1	-	-	Softmax

Table 4. Result of different test tasks.

Tasks	Source Domain	Target Domain	Accuracy (%)
T1	L0	L30	100.00%
T2	L0	L70	98.98%
T3	L30	L70	100.00%
T4	L70	L0	100.00%
T5	L0	L100	96.67%

4.3. Methods Comparison and Results Discussion

In this section, four methods are employed for comparison on the test tasks T1~T5 to illustrate the superiority of the proposed FDACNN. The details of compared methods are described as below.

The standard DANN method is utilized for comparisons [35]. Training data are sequences that convert fusion data into one dimension.

Another popular domain-adaptation method based on maximum mean discrepancy (DA-MMD) is applied for comparisons [36]. The train data and network structure used in DA-MMD are consistent with the proposed method.

The DACNN model was trained using single vibration signals (DACNN-SV). This model is trained using frequency domain signals and CS2 sequence (2 × 64 × 32 formats).

The DACNN model was trained using single infrared thermal image (DACNN-SI). This model is trained using RGB 3-channels of infrared thermal images (3 × 64 × 32 formats).

The results of different comparison methods are listed in Table 5. Meanwhile, in order to compare the results more intuitively, the bar diagram of results is shown in Figure 5. It can be observed that DACNN_SV has the lowest accuracy in five tasks: 47.38%, 40.88%, 67.44%, 35.56% and 35.44%, respectively. This is due to the fact that the vibration signal is not sensitive to two non-structural faults, which makes it impossible to classify, and the resulting accuracy is low. DACNN_SI has a good performance on T1 and T3, but it is not satisfactory in other test tasks; in particular, the accuracy of T5 is only 83%. It illustrates that infrared thermal images can be used to identify structural and non-structural failure states effectively, but they are susceptible to environmental interference. From the results of DANN and DA-MMD, it can be observed that the accuracy in all test tasks is lower than the proposed FDACNN. This is because two-dimensional data fusion can effectively maintain the fault information contained in the infrared image and vibration signal. Meanwhile, the adversarial domain adaptation network can enable the extracted extractor to extract target domain features that are easy to distinguish.

Table 5. The results of different comparisons methods.

Tasks	DANN	DA-MMD	DACNN_SV	DACNN_SI	Proposed FDACNN
T1	98.98%	100.00%	47.38%	100%	100.00%
T2	92.96%	97.12%	40.88%	95.69%	98.98%
T3	97.38%	98.97%	67.44%	100%	100.00%
T4	80.00%	92.69%	35.56%	97.94%	100.00%
T5	60.00%	85%	35.44%	83%	96.67%

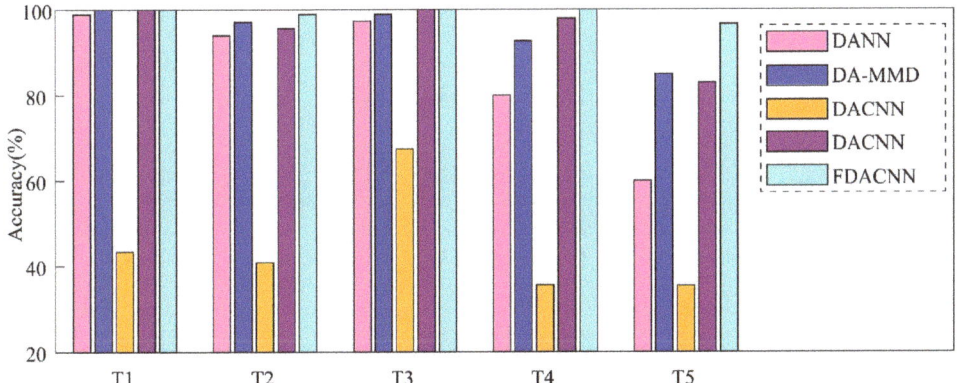

Figure 5. Bar diagram of results in different test tasks.

Additionally, in order to demonstrate the ability of the feature extractor to extract domain invariant features, principal analysis (PCA) is used to map extracted features into 2-dimensional space. Figure 6 shows 2-dimensional visualizations in different test tasks, in which PCA 1 and PCA 2 denote first and second principal components, respectively. It can be observed that the points with the same color are clustered in T1 and T3, and the point clusters of different colors are obviously isolated. In T2, T4 and T5, only a few points with the same color are confused, and most of them with the same color are relatively concentrated. Therefore, extracted features are relatively separable in all test tasks. It suggests that the trained feature extractor has the ability to extract distinguishable features from the unlabeled target domain fusion samples.

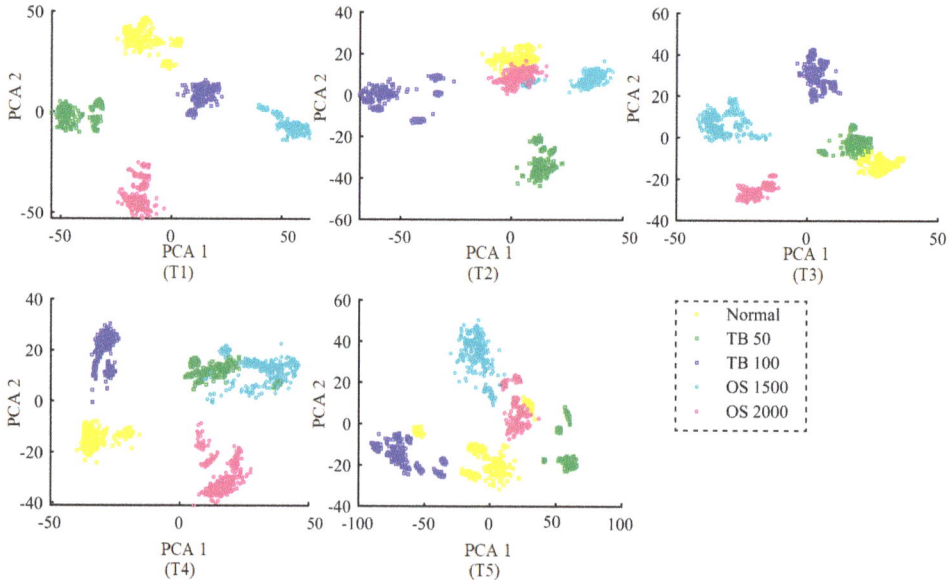

Figure 6. Feature visualization of different test tasks.

5. Conclusions

This study focuses on cross-domain fault diagnosis of gearboxes via multi-source heterogeneous data. Infrared thermal images and vibration signals are fused to characterize the health states of the gearbox, which can effectively recognize structural and non-structural faults. Moreover, the domain-adaptation neural network is trained via adversarial training using fusion data samples to extract the common transfer knowledge, called the FDACNN method. By performing this, the proposed FDACNN method can be used to recognize the unlabeled target domain samples of gearbox. For validation, the proposed FDACNN method is used to analyze gearbox multi-source heterogeneous data measured under various operating conditions. Moreover, we compare the FDACNN method with four other relevant methods to confirm its superiority in cross-domain fault diagnosis of gearboxes under various operating conditions. The results demonstrate that the proposed method obtains highest classification accuracy among four methods.

Author Contributions: Data curation, Z.Z.; Methodology, Y.L.; Writing—original draft, G.M.; Writing—review & editing, B.Q. All authors have read and agreed to the published version of the manuscript.

Funding: The research is supported by National Natural Science Foundation of China under Grant 51805434 and 12172290 and Key Laboratory of Equipment Research Foundation under Grant 6142003190208.

Institutional Review Board Statement: Not applicable.

Informed Consent Statement: Not applicable.

Data Availability Statement: The data presented in this study are available on request from the corresponding author.

Conflicts of Interest: The authors declare that they have no conflict of interest.

References

1. Kong, X.; Mao, G.; Wang, Q.; Ma, H.; Yang, W. A multi-ensemble method based on deep auto-encoders for fault diagnosis of rolling bearings. *Measurement* **2020**, *151*, 107132. [CrossRef]
2. Karabacak, Y.E.; Özmen, N.G.; Gümüşel, L. Intelligent worm gearbox fault diagnosis under various working conditions using vibration, sound and thermal features. *Appl. Acoust.* **2022**, *186*, 108463. [CrossRef]
3. Miao, Y.; Zhang, B.; Lin, J.; Zhao, M.; Liu, H.; Liu, Z.; Li, H. A review on the application of blind deconvolution in machinery fault diagnosis. *Mech. Syst. Signal Process.* **2022**, *163*, 108202. [CrossRef]
4. Shi, J.; Peng, D.; Peng, Z.; Zhang, Z.; Goebel, K.; Wu, D. Planetary gearbox fault diagnosis using bidirectional-convolutional LSTM networks. *Mech. Syst. Signal Process.* **2022**, *162*, 107996. [CrossRef]
5. Liu, H.; Zhou, J.; Zheng, Y.; Jiang, W.; Zhang, Y. Fault diagnosis of rolling bearings with recurrent neural network-based autoencoders. *ISA Trans.* **2018**, *77*, 167–178. [CrossRef] [PubMed]
6. Zhang, W.; Li, C.; Peng, G.; Chen, Y.; Zhang, Z. A deep convolutional neural network with new training methods for bearing fault diagnosis under noisy environment and different working load. *Mech. Syst. Signal Process.* **2018**, *100*, 439–453. [CrossRef]
7. Yu, J.; Liu, X. One-dimensional residual convolutional auto-encoder for fault detection in complex industrial processes. *Int. J. Prod. Res.* **2021**, *196*, 1–20. [CrossRef]
8. Li, D.; Zhang, M.; Kang, T.; Li, B.; Xiang, H.; Wang, K.; Pei, Z.; Tang, X.; Wang, P. Fault diagnosis of rotating machinery based on dual convolutional-capsule network (DC-CN). *Measurement* **2022**, *187*, 110258. [CrossRef]
9. Zhu, J.; Chen, N.; Peng, W. Estimation of bearing remaining useful life based on multiscale convolutional neural network. *IEEE Trans. Ind. Electron.* **2019**, *66*, 3208–3216. [CrossRef]
10. Yao, D.; Liu, H.; Yang, J.; Li, X. A lightweight neural network with strong robustness for bearing fault diagnosis. *Measurement* **2020**, *159*, 107756. [CrossRef]
11. Shao, H.; Jiang, H.; Lin, Y.; Li, X. A novel method for intelligent fault diagnosis of rolling bearings using ensemble deep auto-encoders. *Mech. Syst. Signal Process.* **2018**, *102*, 278–297. [CrossRef]
12. Guo, L.; Lei, Y.; Xing, S.; Yan, T.; Li, N. Deep Convolutional Transfer Learning Network: A New Method for Intelligent Fault Diagnosis of Machines with Unlabeled Data. *IEEE Trans. Ind. Electron.* **2019**, *66*, 7316–7325. [CrossRef]
13. Han, T.; Liu, C.; Yang, W.; Jiang, D. A novel adversarial learning framework in deep convolutional neural network for intelligent diagnosis of mechanical faults. *Knowl. Based Syst.* **2019**, *165*, 474–487. [CrossRef]
14. Qian, W.; Li, S.; Wang, J.; Xin, Y.; Ma, H. A New Deep Transfer Learning Network for Fault Diagnosis of Rotating Machine Under Variable Working Conditions. In Proceedings of the 2018 Prognostics and System Health Management Conference (PHM-Chongqing), Chongqing, China, 26–28 October 2018; pp. 1010–1016.
15. Chen, Z.; Gryllias, K.; Li, W. Intelligent Fault Diagnosis for Rotary Machinery Using Transferable Convolutional Neural Network. *IEEE Trans. Ind. Inform.* **2019**, *16*, 339–349. [CrossRef]
16. Yang, B.; Lei, Y.; Jia, F.; Xing, S. An intelligent fault diagnosis approach based on transfer learning from laboratory bearings to locomotive bearings. *Mech. Syst. Signal Process.* **2019**, *122*, 692–706. [CrossRef]
17. Xu, K.; Li, S.; Wang, J.; An, Z.; Qian, W.; Ma, H. A novel convolutional transfer feature discrimination network for imbalanced fault diagnosis under variable rotational speed. *Meas. Sci. Technol.* **2019**, *30*, 105107. [CrossRef]
18. Wen, L.; Gao, L.; Li, X. A new deep transfer learning based on sparse auto-encoder for fault diagnosis. *IEEE Trans. Syst. Man Cybern.-Syst.* **2017**, *49*, 136–144. [CrossRef]
19. Zhang, M.; Wang, D.; Lu, W.; Yang, J.; Li, Z.; Liang, B. A Deep Transfer Model With Wasserstein Distance Guided Multi-Adversarial Networks for Bearing Fault Diagnosis under Different Working Conditions. *IEEE Access* **2019**, *7*, 65303–65318. [CrossRef]
20. Zhang, Z.; Li, X.; Wen, L.; Gao, L.; Gao, Y. Fault Diagnosis Using Unsupervised Transfer Learning Based on Adversarial Network. In Proceedings of the 2019 IEEE 15th International Conference on Automation Science and Engineering (CASE), Vancouver, BC, Canada, 22–26 August 2019; pp. 305–310.
21. Zhang, B.; Li, W.; Hao, J.; Li, X.-L.; Zhang, M. Adversarial adaptive 1-D convolutional neural networks for bearing fault diagnosis under varying working condition. *arXiv* **2018**, arXiv:1805.00778.
22. Wang, B.; Shen, C.; Yu, C.; Yang, Y. Data Fused Motor Fault Identification Based on Adversarial Auto-Encoder. In Proceedings of the 2019 IEEE 10th International Symposium on Power Electronics for Distributed Generation Systems (PEDG), Xi'an, China, 3–6 June 2019; pp. 299–305.
23. Yang, S.; Kong, X.; Wang, Q.; Li, Z.; Cheng, H.; Yu, L. A multi-source ensemble domain adaptation method for rotary machine fault diagnosis. *Measurement* **2021**, *186*, 110213. [CrossRef]
24. He, Z.; Shao, H.; Zhong, X.; Zhao, X. Ensemble transfer CNNs driven by multi-channel signals for fault diagnosis of rotating machinery cross working conditions. *Knowl. Based Syst.* **2020**, *207*, 106396. [CrossRef]
25. Shi, Y.; Deng, A.; Ding, X.; Zhang, S.; Xu, S.; Li, J. Multisource domain factorization network for cross-domain fault diagnosis of rotating machinery: An unsupervised multisource domain adaptation method. *Mech. Syst. Signal Process.* **2022**, *164*, 108219. [CrossRef]
26. Xin, L.; Haidong, S.; Hongkai, J.; Jiawei, X. Modified Gaussian convolutional deep belief network and infrared thermal imaging for intelligent fault diagnosis of rotor-bearing system under time-varying speeds. *Struct. Health Monit.* **2021**, *99*, 8957. [CrossRef]

27. Choudhary, A.; Mian, T.; Fatima, S. Convolutional neural network based bearing fault diagnosis of rotating machine using thermal images. *Measurement* **2021**, *176*, 109196. [CrossRef]
28. Jia, Z.; Liu, Z.; Vong, C.-M.; Pecht, M. A rotating machinery fault diagnosis method based on feature learning of thermal images. *IEEE Access* **2019**, *7*, 12348–12359. [CrossRef]
29. Bai, T.; Yang, J.; Yao, D.; Wang, Y. Information Fusion of Infrared Images and Vibration Signals for Coupling Fault Diagnosis of Rotating Machinery. *Shock. Vib.* **2021**, *2021*, 6622041. [CrossRef]
30. Di, Z.; Shao, H.; Xiang, J. Ensemble deep transfer learning driven by multisensor signals for the fault diagnosis of bevel-gear cross-operation conditions. *Sci. China Technol. Sci.* **2021**, *64*, 481–492. [CrossRef]
31. Xu, L.; Chatterton, S.; Pennacchi, P. Rolling element bearing diagnosis based on singular value decomposition and composite squared envelope spectrum. *Mech. Syst. Signal Process.* **2021**, *148*, 107174. [CrossRef]
32. Kia, S.H.; Henao, H.; Capolino, G.-A. Efficient digital signal processing techniques for induction machines fault diagnosis. In Proceedings of the 2013 IEEE Workshop on Electrical Machines Design, Control and Diagnosis (WEMDCD), Paris, France, 11–12 March 2013; pp. 232–246.
33. Li, Y.; Gu, J.X.; Zhen, D.; Xu, M.; Ball, A. An evaluation of gearbox condition monitoring using infrared thermal images applied with convolutional neural networks. *Sensors* **2019**, *19*, 2205. [CrossRef]
34. Yongbo, L.; Xiaoqiang, D.; Fangyi, W.; Xianzhi, W.; Huangchao, Y. Rotating machinery fault diagnosis based on convolutional neural network and infrared thermal imaging. *Chin. J. Aeronaut.* **2020**, *33*, 427–438.
35. Ganin, Y.; Ustinova, E.; Ajakan, H.; Germain, P.; Larochelle, H.; Laviolette, F.; Marchand, M.; Lempitsky, V. Domain-adversarial training of neural networks. *J. Mach. Learn. Res.* **2016**, *17*, 2030–2096.
36. Lu, W.; Liang, B.; Cheng, Y.; Meng, D.; Yang, J.; Zhang, T. Deep model based domain adaptation for fault diagnosis. *IEEE Trans. Ind. Electron.* **2016**, *64*, 2296–2305. [CrossRef]

Article

A Study on Railway Surface Defects Detection Based on Machine Vision

Tangbo Bai [1,2,*], Jialin Gao [1,2], Jianwei Yang [1,2] and Dechen Yao [1,2]

1. School of Mechanical-Electronic and Vehicle Engineering, Beijing University of Civil Engineering and Architecture, Beijing 100044, China; 2108230420003@stu.bucea.edu.cn (J.G.); yangjianwei@bucea.edu.cn (J.Y.); yaodechen@bucea.edu.cn (D.Y.)
2. Beijing Key Laboratory of Performance Guarantee on Urban Rail Transit Vehicles, Beijing University of Civil Engineering and Architecture, Beijing 100044, China
* Correspondence: baitangbo@bucea.edu.cn

Abstract: The detection of rail surface defects is an important tool to ensure the safe operation of rail transit. Due to the complex diversity of track surface defect features and the small size of the defect area, it is difficult to obtain satisfying detection results by traditional machine vision methods. The existing deep learning-based methods have the problems of large model sizes, excessive parameters, low accuracy and slow speed. Therefore, this paper proposes a new method based on an improved YOLOv4 (You Only Look Once, YOLO) for railway surface defect detection. In this method, MobileNetv3 is used as the backbone network of YOLOv4 to extract image features, and at the same time, deep separable convolution is applied on the PANet layer in YOLOv4, which realizes the lightweight network and real-time detection of the railway surface. The test results show that, compared with YOLOv4, the study can reduce the amount of the parameters by 78.04%, speed up the detection by 10.36 frames per second and decrease the model volume by 78%. Compared with other methods, the proposed method can achieve a higher detection accuracy, making it suitable for the fast and accurate detection of railway surface defects.

Keywords: deep learning; rail surface defect detection; machine vision; YOLOv4; MobileNetV3

1. Introduction

With the prosperous development of the railway industry, the mileage, speed and density of operations continue to increase, and the inspection requirements for railways are further improved [1]. When it runs at high speed, the phenomena such as friction, rolling contact and elastic deformation occur between the train and the track surface. With the running time increasing, it will result in rail surface defects, such as rail wear, broken, peeling and cracks, which seriously threaten the safety of the rail transit system [2]. Therefore, it is particularly important to study the detection methods for railway surface defects.

As a traditional method for rail surface detection, manual inspection [3] is characteristic of time-consuming, labor-intensive [4] and low detection efficiency [5]. With the development of defect detection technology, many rail surface defect detection methods have emerged, such as ultrasonic flaw detection [6], eddy current flaw detection [7], three-dimensional detection [8], radar detection [9] and so on. The above methods are very effective in detecting internal defects. However, the signals generated by the defects on railway surfaces are very weak, and they are difficult to detect by the above methods. At the same time, the defect signals are easily interfered with by the surrounding environment, leading it difficult to achieve satisfying results. There is still a big margin for improvement in the detection technology of rail surface defects.

With the development of computer technology, the machine vision [10] method is applied to rail surface defect detection. Rail surface detection images are obtained by linear array cameras, and the images are automatically synthesized according to the required length. Defect data are obtained by manual screening from actual detection images for model training and testing. This method requires an analysis of rail surface defect information, gray information [11] and background information [12]. It needs to use a feature extraction algorithm [13] or to use an operator template and model-based threshold segmentation method [14] to detect rail surface defects. However, these methods are susceptible to defect characteristics that may lead to blind spot detection [15]. This makes it difficult for machine vision methods to obtain good detection performances.

In recent years, with the development of target detection technology and the neural network [16], deep learning frameworks have been proposed for the detection of various railway components. Liu et al. [17] proposed a method based on image fusion features and Bayesian compression image classification and recognition, which detected the status of fasteners by extracting improved edge orientation histograms (IEOH) and macroscopic local binary pattern (MSLBP) features. Cui et al. [18] segmented the fastener image into different parts to avoid the interference of the fastener fragments and tested the segmentation model in a real-time deep learning module.

In the application of a deep learning framework for rail surface defect detection, Xu et al. [19] proposed to improve the Faster R-CNN (Convolutional Neural Networks) for railway subgrade defect recognition. The improved method can obtain good performance, but it has disadvantages such as a slow detection speed and large detection model. Lu et al. [20] proposed to apply the combined U-Net graph segmentation network and damage location method for damage detection of high-speed railways. This method can obtain a high detection accuracy but has the limitations of slow detection speed and large model volume. Yuan et al. [21] proposed the application of MobileNetV2 to detect rail surface defects, which achieved high-speed real-time detection, but the detection accuracy was low. Faghih-Roohi et al. [22] proposed improved deep convolutional neural networks (DCNN) to efficiently extract and recognize image features, and a small batch gradient descent method was used to optimize the network for the automatic detection of track surface defects. This method requires a long time for network training. Song et al. [23] proposed a deep learning method where the YOLOv3 (You Only Look Once, YOLO) algorithm was used to detect rail surface defects. This method has a fast detection speed but low detection accuracy.

In order to solve the above problems, this paper proposes an improved YOLOv4 [24] rail surface detection method. It studies the use of the MobileNetV3 lightweight network as the backbone of YOLOv4. Depthwise separable convolution is applied for the PANet layer in YOLOv4 to further reduce the amounts of the parameters. It treats rail surface defect detection as an end-to-end regression problem and ensures the effectiveness of rail surface defect detection with a simplified network, improving the detection speed and accuracy. It provides a new idea for rail surface defect detection technology.

The main contributions of this paper are as follows: (1) The MobileNetV3 network is proposed to optimize the YOLOv4 model for rail surface defect detection, using depthwise separable convolution for the PANet layer in YOLOv4. This method optimizes the parameter quantity and model size and improves the detection speed. (2) Field tests are conducted on the track to collect data, a dataset is created with Gaussian noise added, and finally, a rail surface defect detection model is established. The test results show that the method used in the study can effectively detect rail surface defects.

The rest of this article is organized as follows. The second part discusses the theoretical background of YOLOv4 and depth separable convolution. The third part gives the technical route of the proposed method. The fourth part verifies the effectiveness of the method through practical application. Finally, the conclusion is drawn in Section 5.

2. Theoretical Background

The deep learning and machine vision-based object detection methods are widely used in the current research. For the application of these methods, firstly, a large number of images is collected to establish the image datasets, and secondly, image annotation is performed on the object to be detected in the dataset to obtain the object information; then, a training dataset and the object information are trained by the deep network to obtain a deep network model, and finally, the trained model is used for the object detection test. Among them, the most important part is the training of the deep network model. At this stage, the target detector is mainly composed of four parts: input, backbone, neck and head. As shown in Figure 1, the structure of the one-stage network is simpler than the two-stage one, in which a sparse prediction is added.

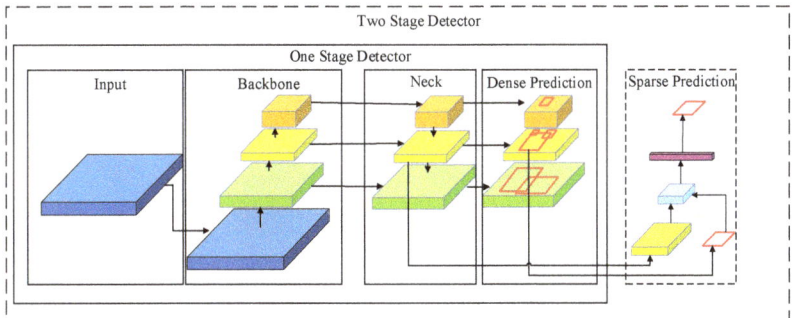

Figure 1. Object detector framework.

Before the YOLO [25] algorithm was proposed, the R-CNN [26] algorithm was one of the most popular algorithms in the two-stage field. CNN has been applied to target detection and formed a relationship with R-CNN [27], the algorithm region. First, the selective search [28] or edge box of the algorithm is used to generate candidate regions [29], and then, each region is trained and classified in the CNN. Compared with the one-stage algorithms, the detection speed of the two-stage ones is slower. Therefore, a YOLO algorithm with the characteristics of the one-stage network structure is proposed. Its core concept is to convert the target detection into a regression problem, and the target map is used as the input of the network. Only through a neural network can the position of the bounding box and the target category be obtained. A fast detection speed and high precision can be realized through the feature information.

The YOLOv4 algorithm is improved from the basis of YOLOv3. As a powerful target detection algorithm, a fast and accurate target detector can be trained by YOLOv4. As shown in Figure 1, the network structure is mainly composed of a backbone network, a neck network and a head network. CSPDarknet53 is applied in the backbone network, an SPP add-on module and PANet path aggregation is performed in the neck network and the YOLOv3 head network is used as the head network.

The PANet layer uses an instance segmentation algorithm. The network structure is shown in the neck part of Figure 2. Compared with the feature pyramid networks (FPN) network, the DownSample operation is added in PANet after UpSample to repeatedly improve the features. Parameter aggregation is carried out on the different backbone layers. It further improves the ability of feature extraction. In YOLOv4, the PANet structure is mainly used in the three effective feature layers.

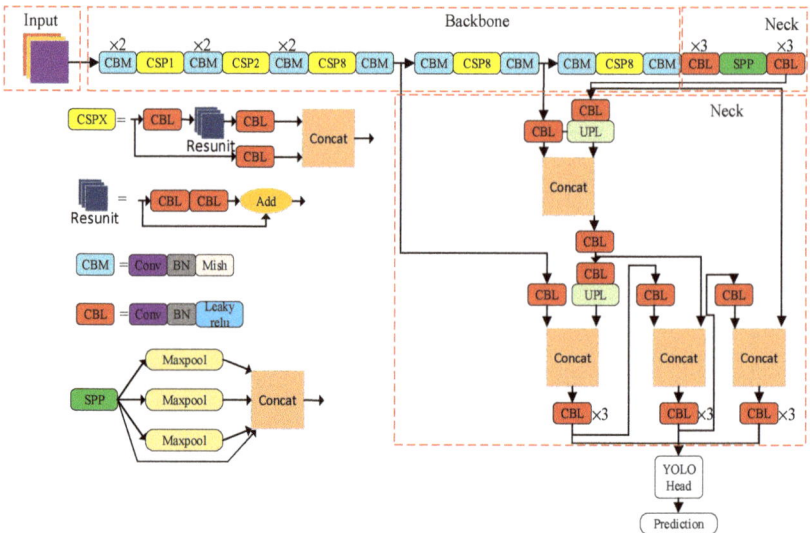

Figure 2. YOLOv4 structure diagram.

3. Proposed Method

3.1. Technical Route

Figure 3 shows the technical route of rail surface defect detection. Firstly, feature extraction is performed on the whole rail image. While retaining the rail surface information, the invalid information is removed from the rail image to increase the network training speed. Secondly, the processed rail surface dataset is input into the improved YOLOv4 network for training. Then, the trained model is used to predict the rail surface defects. Finally, the rail surface defect detection results are obtained.

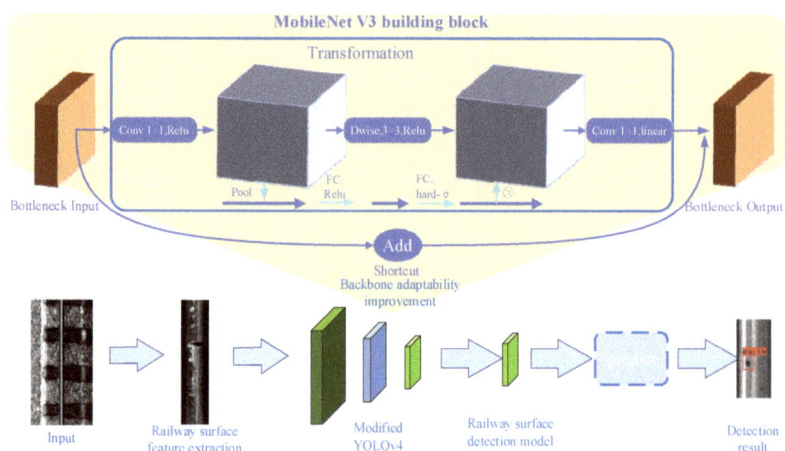

Figure 3. Technical route of the proposed method.

3.2. YOLOv4 Backbone Network Adaptability Improvement

In practical engineering applications, the detection of rail surface defects has particularities, including the accuracy, the speed and the model size of detection. The method in this paper takes into account the particularity of rail surface defect detection, making it adaptable to YOLOv4. MobileNetV3 is used as the backbone network of YOLOv4. MobileNet is a lightweight deep neural network proposed by Google for embedded devices. The core idea is the depthwise separable convolution. Compared with the traditional convolution used in YOLOv4, the deep separable convolution in MobileNetV3 can further reduce the amount of parameters and calculations, thus realizing the lightweight of the network.

A lightweight attention (Squeeze-and-Excitation, SE) module is used in MobileNetV3. Its advantage is that it can improve the performance of the algorithm with a negligible increase in the calculations. The specific process of the SE module is implemented as shown in Figure 4. First, the features of $C' \times H' \times W'$ are optimized to $C \times H \times W$. Then, in the process of squeeze, global average pooling is performed on the $C \times H \times W$ features to obtain a global receptive field feature map of $1 \times 1 \times C$ in size. Then, a fully connected neural network is used for nonlinear transformation in the process of excitation. Finally, the input feature is weighted by the activation value of each feature layer from the SE module.

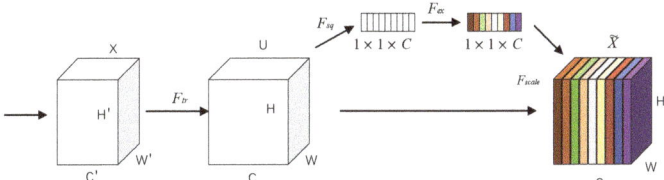

Figure 4. The RPN candidate box generation process.

3.3. Adaptability Improvement of PANet Layer in YOLOv4

PANet in YOLOv4 has the advantages of dynamic feature pooling, fully connected layer fusion and bottom-up path enhancement but disadvantages such as a large amount of parameters and complex calculations. To resolve this problem, the convolution structure in PANet is modified, where the 3×3 and 5×5 standard convolutions are replaced by depth separable convolutions.

Depth separable convolution [30] is a lightweight convolution module. It consists of the following two parts: depthwise convolution (DW) and pointwise convolution (PW). In DW, each dimension in the input information is convolved with a convolution block separately. Then, PW applies a point convolution kernel to perform dimensional lifting of the output maps from DW.

In the standard convolutional layer, assume that the size of the input feature map is $D_z \times D_z$, the number of channels is M, the size of the convolution kernel is $D_i \times D_i$ and the number of convolution kernels is K. Then, the standard convolution calculation amount C_1 can be calculated by Formula (1):

$$C_1 = D_z \times D_z \times M \times K \times D_i \times D_i \tag{1}$$

In depth separable convolution, DW and PW are performed separately, as shown in Figure 5. The calculation amount C_2 of the depth separable convolution can be calculated as Formula (2):

$$C_2 = D_z \times D_z \times M \times D_i \times D_i + K \times M \times D_z \times D_z \tag{2}$$

The calculation amounts of the depth separable convolution and classic convolution are compared as follows:

$$\frac{C_2}{C_1} = \frac{D_z \times D_z \times M \times D_i \times D_i + K \times M \times D_z \times D_z}{D_z \times D_z \times M \times K \times D_i \times D_i} = \frac{1}{K} + \frac{1}{D_i^2} \tag{3}$$

In the equation, the channels number of the convolutional layer K is usually greater than 1, and the commonly used sizes of the convolution kernel are 3×3 and 5×5, so the result of the formula is less than 1. The calculation amount of the depth separable convolution is smaller than that of the standard convolution.

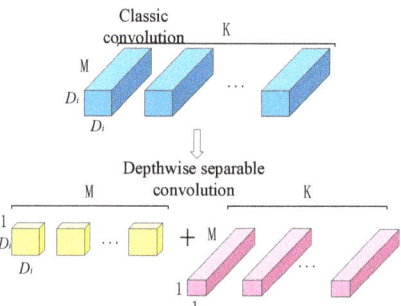

Figure 5. Classic convolution and depth separable convolution.

The PANet layer is improved, as shown in Figure 6. It can retain the advantages of PANet dynamic feature pooling, fully connected layer fusion and bottom-up path enhancement and also reduce the computation in PANet, so as to realize the lightweight of the network and, finally, achieve the optimization of YOLOv4.

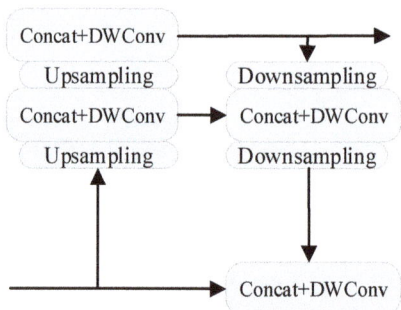

Figure 6. Improved PANet layer.

4. Case Studies

4.1. Image Acquisition

According to the technical route of the proposed method, a track inspection field test was carried out in this paper. As shown in Figure 7, the intelligent track inspection vehicle used in the test was developed by Beijing Yinglu Technology Co., Ltd. (Beijing, China). The vehicle is composed of two parts: an electric inspection vehicle and a track state inspection system. The electric inspection vehicle contains a car body, track wheels and seats; the track state inspection system is composed of a host and a high-definition linear image scanning module. In this test, a 15 km track on the Beijing–Shanghai high-speed rail line is chosen as the test section. The travel speed of the inspection vehicle is 20 km/h, and the image resolution is 2048 × 2048.

Figure 7. Intelligent track inspection vehicle.

The specific collection equipment data is shown in Table 1.

Table 1. Track inspection vehicle camera parameters.

Camera Model	TVI-LCM-01
	2-K linear array image acquisition module
Voltage input range	20–30 V DC
power	120 W
Protection class	
Working temperature	$-20\ °C$ to $+70\ °C$
storage temperature	$-40\ °C$ to $+85\ °C$

The specific configuration of the algorithm environment used in the test is shown in Table 2.

Table 2. Test environment.

Project	Environment
Development language	Python 3.9
Development framework	PyTorch1.2
CPU	Intel(R) i7-9700 CPU @ 3.00 GHz
GPU	NVIDIA GeForce RTX 2080 Ti
Running memory	16 GB
Hard disk size	1 TB

One thousand rail images collected in the field test are chosen for rail surface defect detection; among which, 900 are randomly selected as the training dataset and 100 as the test dataset. Before applying the improved YOLOv4, it needs to operate image annotation to establish a dataset feature database. In this paper, LABELIMG software with version 1.0 was used for image annotation. LABELIMG is an image annotation tool that is written in python and uses QT as a graphical interface. The rail surface in the image is regarded as the target detection area, as shown in Figure 8.

After annotation, the coordinates of the rail surface defect area are obtained, and the training algorithm and the defect detection test are performed on the coordinate dataset generated by the image annotation.

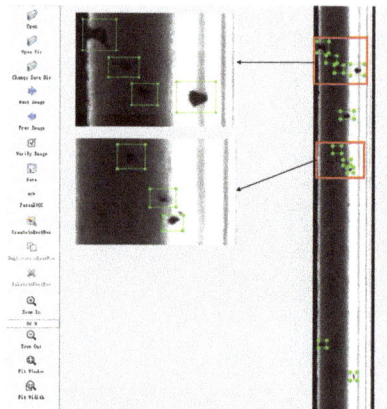

Figure 8. Image annotation.

4.2. Establish a Detection Model for Rail Surface Defects

In order to verify the effectiveness of the method proposed in the study, 5% and 10% Gaussian noise are added to the original dataset, respectively, as shown in Figure 9.

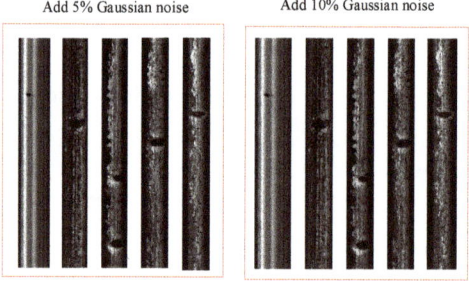

Figure 9. Gaussian noise processing diagram.

The improved YOLOv4 uses MobileNetV3 as the backbone network of the feature extraction and, at the same time, uses deep separable convolution to replace the traditional convolution in PANet. A rail defect detection model is established, as shown in Figure 10, (1) to reset the size of the input image, (2) to apply the improved YOLOv4 network based on the image operation and (3) to output the detection target.

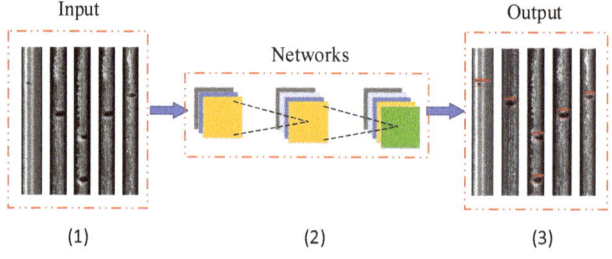

Figure 10. Rail defect detection model. (1) to reset the size of the input image; (2) to apply the improved YOLOv4 network based on the image operation; (3) to output the detection target.

The *CIOU* calculation method in YOLOv4 will make the target frame regression stable. It takes into account the distance, overlap, scale and penalty items between the target and the anchor point, and there will be no training divergence problem. Figure 11 illustrates the surface defects of the rail, and the red box indicates the target frame in which the rail surface defects are surrounded. The green box is the prediction box, and the purple box is the smallest rectangle that can cover the above two. d represents the center point distance between the target box and the predicated one. c represents the diagonal distance of the smallest area simultaneously covering the prediction box and the target box.

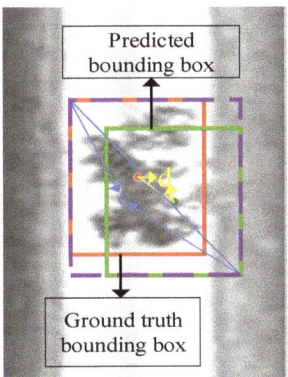

Figure 11. Comparison of the parameters of the methods.

CIOU calculation formula is as shown in Formulas (4)–(6):

$$v = \frac{4}{\pi^2}\left(\arctan\frac{w^{gt}}{h^{gt}} - \arctan\frac{w}{h}\right)^2 \tag{4}$$

$$\alpha = \frac{v}{1 - IOU + v} \tag{5}$$

$$CIOU = IOU - \frac{\rho^2\left(b, b^{gt}\right)}{c^2} - \alpha v \tag{6}$$

where ρ refers to Euclidean distance; b, w and h refer to the center coordinates, width and height of the prediction box and b^{gt}, w^{gt} and h^{gt} refer to the center coordinates, width and height of the frame.

In the study, the *CIOU* threshold was set to 0.7. The detection image can be output only when the result is greater than 0.7, which makes the bounding box more accurate.

In the establishment of the rail defect detection model, the learning rate and the step size for each update is too large; thus, the model cannot converge on the extreme optimal value. If the learning rate is too small, the convergence can be guaranteed, but the efficiency of the model is sacrificed.

In order to avoid the above-mentioned problems, trade-offs have to be considered by modifying the model parameters with the best performances. The adaptive learning rate is used in the experiment to improve the optimization speed of the model, and the initial value of the learning rate is set to 0.001. In the training process, after each epoch, the current model loss and accuracy are evaluated in the training set, and the loss value change is detected every other epoch. When it is less than 0.0001, the learning rate *lr* is attenuated. The attenuation formula is expressed as Formula (7):

$$lr^* = lr \times 0.1 \tag{7}$$

4.3. Result Analysis

In the study, the same dataset is applied on the Faster R-CNN, YOLOv3 and YOLOv4 methods to compare and verify the effectiveness of the proposed method.

Figure 12 illustrates the comparison of the parameter quantity of each method. It shows that the parameter quantity in the proposed method is the least, which is about 1/20 of the Faster R-CNN. Since YOLOv4 is improved from the basis of YOLOv3, the parameter quantities of the two are not much different. Improved from the basis of YOLOv4, the proposed method replaces lightweight MobileNetv3 as the backbone network and uses deep separable convolution for PANet to further reduce the amounts of the parameters. From Table 3, the parameter quantity in the proposed method is decreased by 78.04% compared with YOLOv4, effectively reducing the amounts of the parameters.

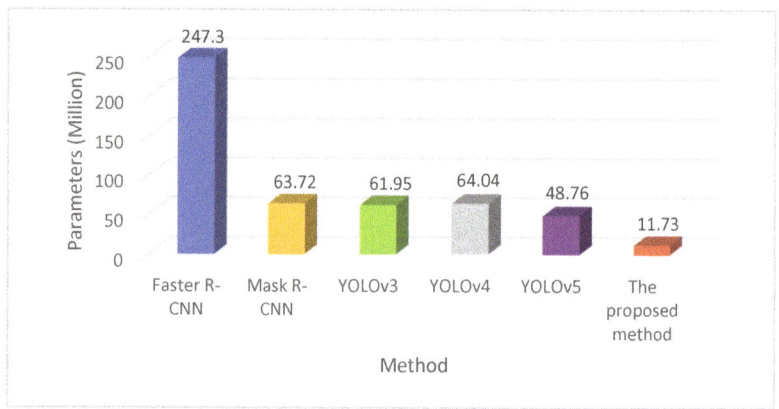

Figure 12. Comparison of the parameter quantities of the methods.

Table 3. Comparison of the detection results of rail defects.

Method	P_r	R_e	mAP	FPS (Hz)	Volume (MB)
Faster R-CNN	89.36%	79.07%	87.32%	12.26	521.8
Mask R-CNN	90.62%	81.36%	89.18%	5.60	245.4
YOLOv3	87.23%	77.27%	86.74%	28.40	234.2
YOLOv4	92.48%	81.40%	90.98%	34.28	244.1
YOLOv5	93.06%	82.08%	92.16%	37.32	185.6
The proposed method	94.24%	82.56%	93.21%	44.64	53.6

In order to evaluate the detection results of rail defects, precision (P_r), recall (R_e), mean Average Precision (mAP), Frames Per Second (FPS) and volume are introduced. Among them, mAP is a common parameter for accuracy evaluations of different target detection models. Specifically, it is the mean of the average precision (AP) of each query. FPS refers to the number of frames transmitted per second, and volume refers to the size of the memory occupied by the model. The specific calculation formula is as follows:

$$P_r = \frac{TP}{TP+FP} \times 100\% \tag{8}$$

$$R_e = \frac{TP}{TP+FN} \times 100\% \tag{9}$$

$$AP = \int_0^1 p(r)dr \tag{10}$$

$$mAP = \frac{1}{N}\sum AP_i \tag{11}$$

where True Positives (TP) and False Positives (FP) are the number of rail defects detected correctly or not, respectively. False Negatives (FN) is the number of rail defects detected incorrectly. N is the number of defects in all the rails.

From the detection results of rail surface defects by various methods in Table 3, it can be seen that, as a popular traditional method in the two-stage field, Faster R-CNN has a higher accuracy, recall and mAP than YOLOv3 but a lower detection speed. A slow, large model size not suitable for lightweight real-time detection, YOLOv3 has the advantage of a faster detection speed and smaller model size, but its accuracy, recall rate and mAP and Faster R-CNN methods are small; YOLOv4 in the accuracy, recall rate, mAP and FPS ahead of Faster R-CNN and YOLOv3, and its detection speed and model volume still have room for improvement. The research method in this paper was improved from the basis of YOLOv4. Due to the use of lightweight MobileNet V3 as the backbone network and deep separable convolution to improve the PANet, the model volume was 0.22 times that of YOLOv4, and the accuracy was improved by 1.64% compared to YOLOv4. Compared with YOLOv4, the recall rate and mAP were increased by 1.16% and 2.54%, respectively. At the same time, the detection speed of the research method exceeded YOLOv4 by 10.36 frames per second, which can better meet the requirement of rapidity.

Due to the complex environment of the rail, the algorithm is required to have a good anti-noise performance. In order to test the noise resistance of the research, Gaussian noise was added into the dataset. Tables 4 and 5 are the detection results of rail defects with 5% and 10% Gaussian noise, respectively. It can be seen the proposed method has a higher mAP than the other methods and has more superior performance when noise exists. As the same models are used with slightly different test data, the FPS and volume of each method are consistent with those in Table 3. The results of Tables 3–5 show that the proposed method in this paper has good performance and can be applied to lightweight steel rail surface defect detection.

Table 4. Comparison of the detection results of rail defects with 5% Gaussian noise.

Method	P_r	R_e	mAP	FPS (Hz)	Volume (MB)
Faster R-CNN	82.61%	75.22%	85.08%	12.31	521.8
Mask R-CNN	88.26%	77.93%	86.92%	5.53	245.4
YOLOv3	80.02%	72.73%	83.10%	27.07	234.2
YOLOv4	90.48%	79.36%	87.23%	35.92	244.1
YOLOv5	91.62%	80.14%	90.08%	38.54	13.6
The proposed method	92.44%	80.27%	88.42%	42.78	53.6

Table 5. Comparison of the detection results of rail defects with 10% Gaussian noise.

Method	P_r	R_e	mAP	FPS (Hz)	Volume (MB)
Faster R-CNN	79.35%	71.52%	80.65%	11.80	521.8
Mask R-CNN	85.48%	72.30%	81.23%	5.47	245.4
YOLOv3	75.36%	68.18%	74.40%	28.33	234.2
YOLOv4	88.89%	72.73%	83.02%	32.35	244.1
YOLOv5	91.62%	80.14%	90.08%	36.00	13.6
The proposed method	89.92%	79.63%	84.28%	43.42	53.6

5. Conclusions

The rapid, accurate and intelligent detection of rail surface defects is of great significance for ensuring the safe operations of railway vehicles. According to the characteristics of rail surface defect detection, a one-stage detection model based on deep learning was constructed for the detection of rail surface defects. Through experimental verification and comparative analysis, the following conclusions were drawn:

(1) In order to reduce the weight of the rail surface defect detection network, the YOLOv4 algorithm was improved. The backbone network of YOLOv4 was optimized, and

the PANet layer in YOLOv4 was lightened and improved. It reduced the algorithm parameters, increased the detection speed and reduced the model size.

(2) In order to solve the problem of small objects detection, the improved YOLOv4 method was used in rail surface defect detection. The test results verified the effectiveness of the method.

(3) Establish training and test datasets and adding Gaussian noise processing to the datasets let us conduct the detection case studies. The analysis results showed that, compared with the traditional detection method, the proposed method had a higher detection accuracy.

In addition to the above conclusions, with the rapid development of object detection methods, the ideas proposed in this paper can be extended to different deep learning networks. At the same time, in order to verify the effectiveness of the proposed method and to avoid introducing more variables, image preprocessing was not introduced in this paper. It can be inferred that the accuracy of the defect detection can be further improved if the image is effectively preprocessed. Finally, if sufficient railway surface defect images can be obtained to establish datasets, statistical tests can be performed to achieve a full statistical analysis of the proposed deep learning approaches.

Author Contributions: All authors conceived and designed the study. Conceptualization, methodology, software, validation and writing—original draft, T.B.; software and visualization, J.G.; validation and investigation, J.Y. and software and validation, D.Y. All authors have read and agreed to the published version of the manuscript.

Funding: This research was funded by the General Project of Scientific Research Program of Beijing Municipal Education Commission under Grant KM202010016003, the National Natural Science Foundation of China under Grant 51975038, and the Natural Science Foundation of Beijing under Grant KZ202010016025.

Institutional Review Board Statement: Not applicable.

Informed Consent Statement: Not applicable.

Data Availability Statement: The data used to support the findings of this study are available from the corresponding author upon request.

Acknowledgments: The authors appreciate the support from the Beijing Key Laboratory of Performance Guarantee on Urban Rail Transit Vehicle for this research.

Conflicts of Interest: The authors declare no conflict of interest.

References

1. Nenov, N.; Dimitrov, E.; Vasilev, V.; Piskulev, P. Sensor system of detecting defects in wheels of railway vehicles running at operational speed. In Proceedings of the 2011 34th International Spring Seminar on Electronics Technology, Tratanska Lomnica, Slovakia, 11–15 May 2011; pp. 577–582.
2. Li, Y.; Trinh, H.; Haas, N.; Otto, C.; Pankanti, S. Rail component detection, optimization, and assessment for automatic rail track inspection. *IEEE Trans. Intell. Transp. Syst.* **2014**, *15*, 760–770.
3. Molodova, M.; Li, Z.; Nunez, A.; Dollevoet, R. Automatic detection of squats in railway infrastructure. *IEEE Trans. Intell. Transp. Syst.* **2014**, *15*, 1980–1990. [CrossRef]
4. Badran, W.; Nietlispach, U. Wayside train monitoring systems: Networking for greater safety. *Glob. Railw. Rev.* **2011**, *17*, 14–21.
5. Ho, T.K.; Liu, S.Y.; Ho, Y.T.; Ho, K.H.; Wong, K.K.; Lee, K.Y.; Tam, H.Y.; Ho, S.L. Signature analysis on wheel-rail interaction for rail defect detection. In Proceedings of the 2008 4th IET International Conference on Railway Condition Monitoring, Derby, UK, 18–20 June 2008; pp. 1–6.
6. Clark, R. Rail flaw detection: Overview and needs for future developments. *NDT E Int.* **2004**, *37*, 111–118. [CrossRef]
7. Song, Z.; Yamada, T.; Shitara, H.; Takemura, Y. Detection of damage and crack in railhead by using eddy current testing. *J. Electromagn. Anal. Appl.* **2011**, *3*, 546–550. [CrossRef]
8. Lorente, A.G.; Llorca, D.F.; Velasco, M.G.; García, J.A.R.; Domínguez, F.S. Detection of range-based rail gage and missing rail fasteners: Use of high-resolution two- and three-dimensional images. *Transp. Res. Rec.* **2014**, *2448*, 125–132. [CrossRef]
9. Lohmeier, S.P.; Rajaraman, R.; Ramasami, V.C. Development of an ultra-wideband radar system for vehicle detection at railway crossings. *IEEE Int. Geosci. Remote Sens. Symp.* **2002**, *6*, 3692–3694.

10. Chen, Y.R.; Chao, K.; Kim, M.S. Machine vision technology for agricultural applications. *Comput. Electron. Agric.* **2002**, *36*, 173–191. [CrossRef]
11. Li, Q.; Ren, S. A real-time visual inspection system for discrete surface defects of rail heads. *IEEE Trans. Instrum. Meas.* **2012**, *61*, 2189–2199. [CrossRef]
12. Min, Y.Z.; Yue, B.; Ma, H.F. Rail surface defects detection based on gray scale gradient characteristics of image. *Chin. J. Instrum.* **2018**, *9*, 220–229.
13. Wang, F.; Xu, T.; Tang, T.; Zhou, M.; Wang, H. Bilevel feature extraction-based text mining for fault diagnosis of railway systems. *IEEE Trans. Intell. Transp. Syst.* **2017**, *18*, 49–58. [CrossRef]
14. Banik, P.P.; Saha, R.; Kim, K.D. An Automatic Nucleus Segmentation and CNN model based classification method of white blood cell. *Expert Syst. Appl.* **2020**, *149*, 113211. [CrossRef]
15. Li, Q.; Ren, S. A visual detection system for rail surface defects. *IEEE Trans. Syst. Man Cybern. Part C* **2012**, *42*, 1531–1542. [CrossRef]
16. Meng, Z.; Shi, G.; Wang, F. Vibration response and fault characteristics analysis of gear based on time-varying mesh stiffness. *Mech. Mach. Teory.* **2020**, *148*, 103786. [CrossRef]
17. Liu, J.; Li, B.; Xiong, Y.; He, B.; Li, L. Integrating the symmetry image and improved sparse representation for railway fastener classification and defect recognition. *Math. Probl. Eng.* **2015**, *2015*, 462528. [CrossRef]
18. Cui, H.; Li, J.; Hu, Q.; Mao, Q. Real-time inspection system for ballast railway fasteners based on point cloud deep learning. *IEEE Access* **2020**, *8*, 61604–61614. [CrossRef]
19. Sun, X.; Gu, J.; Huang, R.; Zou, R.; Palomares, B.G. Surface defects recognition of wheel hub based on improved Faster R-CNN. *Electronics* **2019**, *8*, 481. [CrossRef]
20. Lu, J.; Liang, B.; Lei, Q.J.; Li, X.H.; Liu, J.H.; Liu, J.; Xu, J.; Wang, W.J. SCueU-Net: Efficient damage detection method for railway rail. *IEEE Access* **2020**, *8*, 125109–125120. [CrossRef]
21. Yuan, H.; Chen, H.; Liu, S.; Lin, J.; Luo, X. A deep convolutional neural network for detection of rail surface defect. In Proceedings of the 2019 IEEE Vehicle Power and Propulsion Conference, Hanoi, Vietnam, 14–17 October 2019; pp. 1–4.
22. Faghih-Roohi, S.; Hajizadeh, S.; Núñez, A.; Babuska, R.; Schutter, B.D. Deep convolutional neural networks for detection of rail surface defects. In Proceedings of the 2016 International Joint Conference on Neural Networks, Vancouver, BC, Canada, 24–29 July 2016; pp. 2584–2589.
23. Yanan, S.; Hui, Z.; Li, L.; Hang, Z. Rail Surface Defect Detection Method Based on YOLOv3 deep learning networks. In Proceedings of the 2018 Chinese Automation Congress, Xi'an, China, 30 November–2 December 2018; pp. 1563–1568.
24. Bochkovskiy, A.; Wang, C.Y.; Liao, H.Y.M. YOLOv4: Optimal speed and accuracy of object detection. *arXiv* **2020**, arXiv:2004.10934.
25. Redmon, J.; Farhadi, A. YOLOv3: An Incremental Improvement. *arXiv* **2018**, arXiv:1804.02767.
26. Ren, S.; He, K.; Girshick, R.; Sun, J. Faster R-CNN: Towards real-time object detection with region proposal networks. *arXiv* **2016**, arXiv:1506.01497. [CrossRef] [PubMed]
27. Girshick, R.; Donahue, J.; Darrell, T.; Malik, J. Rich feature hierarchies for accurate object detection and semantic segmentation. In Proceedings of the IEEE Conference on Computer Vision and Pattern Recognition, Columbus, OH, USA, 23–28 June 2014; pp. 580–587.
28. Uijlings, J.R.R.; Sande, K.E.A.; Gevers, T.; Smeulders, A.W.M. Selective search for object recognition. *Int. J. Comput. Vis.* **2013**, *104*, 154–171. [CrossRef]
29. Zitnick, C.L.; Dollár, P. Edge boxes: Locating object proposals from edges. In Proceedings of the European Conference on Computer Vision, Zurich, Switzerland, 6–12 September 2014; p. 8693.
30. Chollet, F. Xception: Deep learning with depthwise separable convolutions. In Proceedings of the 2017 IEEE Conference on Computer Vision and Pattern Recognition, Honolulu, HI, USA, 21–26 July 2017; pp. 1800–1807.

Article

A Novel Framework for Anomaly Detection for Satellite Momentum Wheel Based on Optimized SVM and Huffman-Multi-Scale Entropy

Yuqing Li [1], Mingjia Lei [1,*], Pengpeng Liu [2], Rixin Wang [1] and Minqiang Xu [1]

[1] Deep Space Exploration Research Center, Harbin Institute of Technology, Harbin 150080, China; bradley@hit.edu.cn (Y.L.); wangrx@hit.edu.cn (R.W.); xumq@hit.edu.cn (M.X.)
[2] Naval Research Academy, Beijing 100061, China; newtime1987@163.com
* Correspondence: hitleimingjia@163.com

Citation: Li, Y.; Lei, M.; Liu, P.; Wang, R.; Xu, M. A Novel Framework for Anomaly Detection for Satellite Momentum Wheel Based on Optimized SVM and Huffman-Multi-Scale Entropy. *Entropy* **2021**, *23*, 1062. https://doi.org/10.3390/e23081062

Academic Editor: Donald J. Jacobs

Received: 2 July 2021
Accepted: 14 August 2021
Published: 17 August 2021

Publisher's Note: MDPI stays neutral with regard to jurisdictional claims in published maps and institutional affiliations.

Copyright: © 2021 by the authors. Licensee MDPI, Basel, Switzerland. This article is an open access article distributed under the terms and conditions of the Creative Commons Attribution (CC BY) license (https://creativecommons.org/licenses/by/4.0/).

Abstract: The health status of the momentum wheel is vital for a satellite. Recently, research on anomaly detection for satellites has become more and more extensive. Previous research mostly required simulation models for key components. However, the physical models are difficult to construct, and the simulation data does not match the telemetry data in engineering applications. To overcome the above problem, this paper proposes a new anomaly detection framework based on real telemetry data. First, the time-domain and frequency-domain features of the preprocessed telemetry signal are calculated, and the effective features are selected through evaluation. Second, a new Huffman-multi-scale entropy (HMSE) system is proposed, which can effectively improve the discrimination between different data types. Third, this paper adopts a multi-class SVM model based on the directed acyclic graph (DAG) principle and proposes an improved adaptive particle swarm optimization (APSO) method to train the SVM model. The proposed method is applied to anomaly detection for satellite momentum wheel voltage telemetry data. The recognition accuracy and detection rate of the method proposed in this paper can reach 99.60% and 99.87%. Compared with other methods, the proposed method can effectively improve the recognition accuracy and detection rate, and it can also effectively reduce the false alarm rate and the missed alarm rate.

Keywords: satellite momentum wheel; anomaly detection; Huffman-multi-scale entropy (HMSE); support vector machine (SVM); adaptive particle swarm optimization (APSO)

1. Introduction

As important spacecraft, study of the reliability of artificial satellites is a hot topic at present. Generally, an artificial satellite consists of a structural system, temperature control system, attitude control system, measurement and control system, and power supply system. The mission of the attitude control system is to help the satellite achieve attitude stability or attitude maneuver, to further guarantee the normal operation of the satellite platform and the normal work of the payload.

Satellites have high requirements for attitude accuracy, which makes the task of attitude control systems very heavy. Health state and reliability are the basic guarantee for the normal operation of satellites [1]. Therefore, research on the theory and technology of automatic fault diagnosis and anomaly detection of satellite attitude control systems will further ensure the safe and reliable operation of on-orbit aircraft, reducing the possibility of space accidents.

In recent years, many scholars have conducted research on fault diagnosis technology or health management technology. These research contents can be roughly divided into three main aspects. First, when there is a specific research object, a feasible solution is to construct a simulation model of the object by analyzing the working mechanism and failure mode of the object. The data generated based on the simulation data is used as the theoretical prediction value, and then the judgment criterion is designed to complete the detection task. Luo et al. propose an improved phenomenological model based on meshing vibration to generate fault simulation data [2]. Li et al. established an INS/ADS fault detection model based on kinematic equations, and combined an unscented Kalman filter (UKF) with Runge-Kutta to deal with the non-linear and discretization problem [3]. Second, some research aims at extracting the fault features by constructing more effective signal processing methods, such as the feature extraction method based on entropy value [4,5], the feature extraction method based on spectral kurtosis time (Spectral Kurtosis, SK) [6], or the Frequency domain feature extraction method [7]. To fully excavate the features of the momentum wheel telemetry signal, this paper uses a combination of time domain features, frequency domain features and complexity features for feature extraction. Considering that, compared with permutation, dispersion, hierarchy, etc., sample entropy has better consistency for different parameters, this paper chooses a complex quantification method based on sample entropy. Third, for the fault recognition process, various pattern recognition methods are used to learn the mapping relationship between features and failure modes, so as to realize automatic fault recognition [8].

Due to the extremely complex structure and working principle of the spacecraft itself, and the strong coupling between the sub-systems, it is very difficult to construct an accurate simulation model of the spacecraft or its components [9,10]. As the spacecraft is affected by the special space environment during its orbiting operation, it is extremely prone to unpredictable failures, for example, the circuit signal disturbance caused by electromagnetic background [11], the sudden change of attitude caused by the impact of space debris [12], etc. In addition, during the process of the spacecraft downloading telemetry data to the ground-based measurement and control station, data jumps and even partial loss can occur [13]. Therefore, the data generated by the simulation model is often difficult to simulate the actual telemetry data of the spacecraft, and it becomes very difficult to use the spacecraft anomaly detection method based on the physical simulation model in practical applications.

The fault diagnosis method based on the data mode does not impose necessary restrictions on the prior knowledge of the object or system (including mathematical models and expert experience, etc.), such as artificial neural network (ANN), support vector machine (SVM), Bayesian network (BN) and other health assessment methods.

ANN is a method that is widely used in fault identification problems. Multilayer Perceptron (MLP) is the most typical type of feedforward neural network model, which usually uses a BP algorithm to learn the parameters of the model. Kumar et al. proposed a method based on principal component analysis (PCA) and MLP to detect and classify the three-phase current signals online [14]. In addition, probabilistic neural network (PNN) [15], RBF neural network [16], extension neural network (ENN) [17] and recurrent neural network (RNN) [18–20] have also been applied to fault detection and diagnosis problems.

For high-dimensional identification problems in fault diagnosis, the SVM method based on the principle of structural risk minimization has been widely used in recent years [21–23]. Compared with the ANN method based on the principle of empirical risk minimization, the learning goal of the SVM is to learn the optimal classification hyperplane in the feature space. The ANN has the ability to deal with pattern recognition problems, but the sample size is large, and it takes a long time to adjust the network structure parameters. Bayesian decision-making has significant execution ability under the premise of considering prior probability, but good accuracy is based on a prior model with appropriate assumptions. Compared with the above methods, the SVM only needs a small number of samples for training and has better generalization ability. Therefore, this paper chooses SVM as the means of pattern recognition. In the field of fault diagnosis, research on the SVM method mainly focuses on two aspects of obtaining more accurate recognition accuracy, i.e., by optimizing the hyperparameters of the model and constructing a new kernel function. For specific recognition tasks, to optimize the hyperparameters of the model to obtain better recognition performance, many optimization methods are applied [24–26].

Liu et al. proposed a novel small sample data missing filling method based on support vector regression (SVR) and genetic algorithm (GA) to improve the equipment health diagnosis effect [25]. Particle swarm optimization (PSO) is a hyperparameter optimization algorithm which is used by Cuong-Le et al. for damage identifications [26]. In terms of constructing a new kernel function, Wang et al. proposed a kernel function selection mechanism under sparse representation and the superiority of the selection mechanism was performed in simulations and engineering experiments involving high-speed bearing fault diagnosis [27]. Although both GA and PSO can solve high-dimensional complex optimization problems well, in the iterative process of PSO, the particles can retain the memory of the good solution, but the GA cannot, so PSO can often converge to a better solution more quickly. Based on the above analysis, this paper uses PSO to optimize the multi-class SVM.

From the above analysis, it can be seen that the following problems still exist in the direct application of existing anomaly detection or fault diagnosis methods to the anomaly detection problem of the satellite momentum wheel.

(1) Due to the complex structure and control law of the satellite momentum wheel itself, it is very difficult to construct an accurate simulation model, so model-based anomaly detection methods often fail to achieve satellite momentum wheel anomaly detection.

(2) Satellite telemetry data often contains outliers (due to the data with very large deviations introduced by the telemetry process). These data alone cannot characterize the health of the spacecraft, but they can easily be detected as abnormal values by existing methods. At the same time, some segments of the telemetry data are lost in the process of downloading the data from the satellite to the ground. Therefore, reasonable preprocessing of telemetry data is required.

(3) The sampling frequency of telemetry data collected by on-orbit satellite is often less than 1Hz, and the data itself has a long change period, so traditional anomaly detection methods based on time-frequency domain analysis are difficult to work with telemetry data.

Therefore, in response to the above problems, this article proposes a new method based on multi-type features fusion and improved SVM to handle the problem of anomaly detection for the satellite. The main contributions of the proposed framework can be summarized as follows:

(1) We design a new anomaly detection framework for satellites, which includes a telemetry data preprocessing part, a telemetry data multi-type feature extraction part, and a data-driven anomaly detection part.

(2) We propose a new method to construct the fusion-feature sequence HMSE-T/F. The HMSE-T/F is based on the Huffman-multi-scale entropy and the selected time/frequency-domain feature. The Huffman-multi-scale entropy is a new method based on the Huffman coding principle and sample entropy.

(3) We build a multi-class SVM model based on the directed acyclic graph (DAG) principle. We propose an improved adaptive particle swarm optimization (APSO) to train the multi-class SVM model. Compared with other methods, the proposed method has an excellent ability in anomaly detection.

The rest of this paper is organized as follows. Section 2 presents the scheme of the proposed anomaly detection framework. The construction method of multi-type feature sequence HMSE-T/F is provided in Section 3. In Section 4, the anomaly detection method based on multi-class SVM model and the improved adaptive particle swarm optimization (APSO) are stated. In Section 5, the performance of the proposed method is evaluated from different aspects. Finally, in Section 6, a comprehensive summary of this paper and prospects for future work are given.

2. The Scheme of the Proposed Anomaly Detection Framework

2.1. Description of Difficulties in Spacecraft Anomaly Detection

In fact, since satellites are at normal working conditions at most of the time during their orbits, the proportion of normal data in the telemetry data collected on the ground is very high. For most detection methods that rely on plenty of training data, satellite telemetry data can provide very few abnormal or fault samples, and there are very few effective samples that can be used for classification model training. Therefore, some adaptive improvements are needed when using the classification model to detect anomalies in spacecraft.

Figure 1a shows the momentum wheel voltage change of a certain type of satellite within 10 days, and its sampling frequency is 0.125 Hz. Figure 1b shows a sudden voltage change in a certain type of satellite. Figure 1c is the frequency spectrum of the telemetry signal in Figure 1a,d is the partially enlarged view of Figure 1c. According to Figure 1a–d, apart from the feature of less abnormal data, satellite telemetry data also exhibits the characteristics of extremely low sampling frequency, slow data change over a long period of time, and many sudden abnormalities. Therefore, anomaly detection methods that rely on time domain and frequency domain feature extraction often find it difficult to distinguish the health status of their telemetry data.

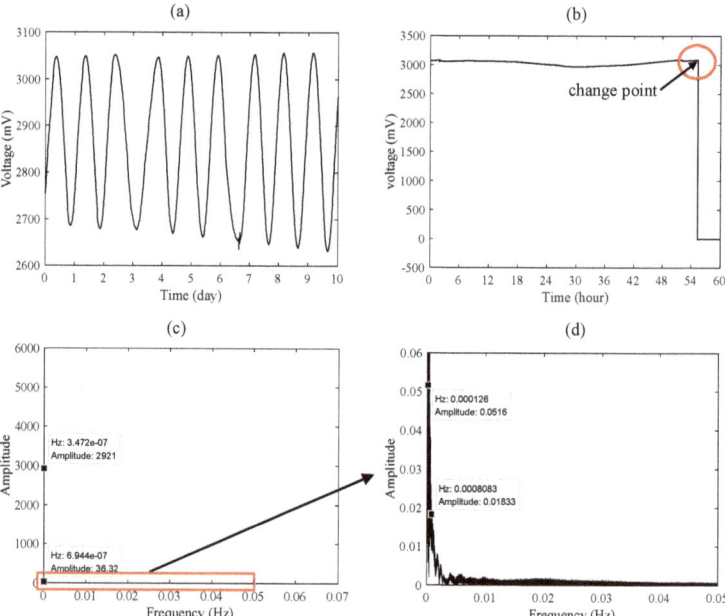

Figure 1. A satellite's momentum wheel voltage telemetry data: (**a**) 10-day data sampled at a frequency of 0.125 Hz, (**b**) sample with sudden change, (**c**) frequency spectrum, (**d**) partially enlarged view of (**c**).

2.2. The Proposed Anomaly Detection Framework

To effectively solve the problem of satellite momentum wheel anomaly detection, a new anomaly detection framework based on multi-type feature extraction and fusion is proposed in this paper. The overall procedure of the proposed anomaly detection framework is shown in Figure 2. Specifically, the descriptions of each Step are detailed as follows.

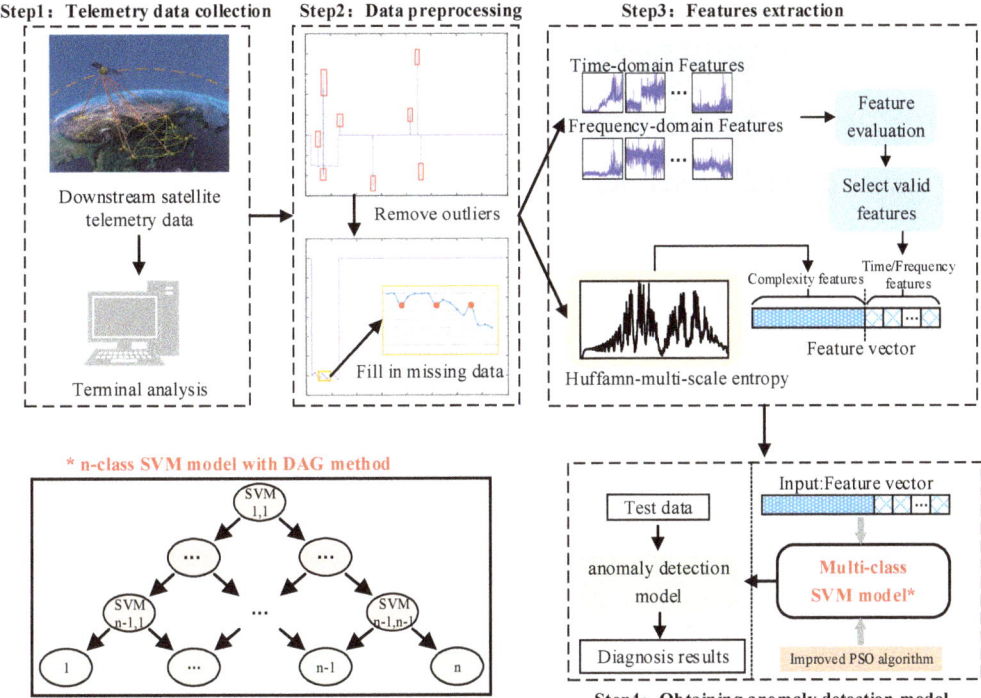

Figure 2. The overall procedure of the proposed anomaly detection framework.

Step 1: Telemetry data collection.

When the satellite is in orbit, to obtain its internal operating status and further provide real-time data for the remote-control object, the sensors in the satellite telemetry system need to measure the operating status of each key component and convert it into electrical signals. After the signals of each channel are combined according to a certain system, they are transmitted to the ground telemetry equipment (including receiver, antenna and splitter demodulator) using radio communication technology, and the ground terminal equipment restores and stores the original parameter information of each channel through signal demodulation technology.

Step 2: Data preprocessing.

The collection process of telemetry data is interfered with by sensors, converters, and wireless transmission. The data obtained by the ground receiving end often produces abnormal jump points. These kind of data points that deviate from the change law of the measured signal are usually called abnormal outliers. The abnormal outliers of the telemetry data will provide wrong information and affect the processing and analysis results of the telemetry signal. Outlier elimination is an important part of telemetry data preprocessing. By eliminating random measurement values with large errors, the authenticity of telemetry data can be guaranteed to a certain extent, and the reliability of data analysis can be improved. Commonly used methods to eliminate outliers include visual inspection, mean square method, point discrimination, Letts criterion, etc. Different outlier elimination methods should be used for different types of telemetry data. Considering that this article mainly analyzes the telemetry data of the satellite momentum wheel, the outlier elimination method based on the Letts criterion is adopted.

The premise of the Letts criterion is that the distribution of the measured data is close to the normal distribution. Based on this assumption, the given confidence probability is 99.7% as the standard, and the standard deviation of three times the measured quantity is used as the basis. Any measurement value exceeding this limit is judged for wild value. For a given sequence of telemetry measurement values. For a given telemetry sequence $x = \{x_i\}, i = 1, \cdots, N$, the specific process of the method is as follows.

(1) Calculate the mean of the series:

$$\overline{x} = \sum_{i=1}^{N} x_i \tag{1}$$

(2) Calculate the standard deviation of the series:

$$\sigma = \frac{1}{N} \sqrt{\sum_{i=1}^{N} (x_i - \overline{x})} \tag{2}$$

(3) Eliminate outliers:

$$\begin{cases} |x_i - \overline{x}| \leq 3\sigma, & \text{not outliers,} \quad \text{keep} \\ |x_i - \overline{x}| > 3\sigma, & \text{outliers,} \quad \text{delete} \end{cases} \tag{3}$$

In addition to the problem of outliers, the process of satellite telemetry data transmission to the ground is affected by the ionosphere, and data may be missing during the signal decoding process. A telemetry sequence that has many data problems should be discarded and not used as training data, but the missing value at a certain point in the sequence can be handled by the filling method. From the distribution of the missing values, they can be divided into missing completely at random (MCAR), missing at random (MAR) and missing not at random (MNAR). MCAR means that the law of missing values in the data is completely random and does not affect the unbiasedness of the overall sample. MAR means that the mechanism of missing data is not completely random. The missing data of this type depends on other variables. Such missing values are relatively rare in telemetry data. MNAR means that the missing data is related to the value of the variable itself.

The missing values in satellite telemetry data are generally MCAR, so this paper uses an interpolation method based on two short sequences before and after the missing point to fill in the missing values. Given the data sequence to be filled is $y = \{y_i\}, i = 1, \cdots, M$, The missing value to be filled is y^*. The auxiliary variable used to construct the regression equation is $x = \{x_j\}, j = 1, \cdots, M$. The auxiliary variable value corresponding to the missing value of the variable to be filled is x^*, and x^* is a known variable. Use x, y to construct the regression equation:

$$y_i = f^*(x_i) \tag{4}$$

where f^* needs to choose different regression models according to different telemetry data. Then the missing value is $y^* = f(x^*)$.

Step 3: Features extraction.

Considering the difficulty of using satellite telemetry data for anomaly detection as mentioned above, this paper adopts a time-frequency domain feature extraction and selection method based on feature quality evaluation. At the same time, a complexity feature extraction method based on Hoffman multi-scale entropy is proposed, which enriches signal feature types and provides effective feature learning samples for training satellite telemetry data anomaly detection models. The specific method of feature extraction is described in detail in Section 3.

Step 4: Obtaining the anomaly detection model.

This paper takes support vector machine (SVM) as the basic unit and uses a directed acyclic graph (DAG) principle to construct a satellite momentum wheel anomaly detection model based on the support vector machine. This model can effectively solve the multi-classification problem when some categories are difficult to distinguish. In addition, to improve the classification accuracy of the anomaly detection model, an improved particle swarm optimization (PSO) algorithm is proposed to train SVMs. The specific method of Obtaining anomaly detection model is described in detail in Section 4.

3. Multi-Type Feature Sequence HMSE-T/F Construction Method

3.1. Time/Frequency Domain Feature Extraction and Selection

3.1.1. Time/Frequency-Domain Feature

The time domain signal is a time series in which time is the independent variable to describe the change of a certain physical quantity, and it is the most basic and most intuitive form of expression of the signal. The time domain signal reflects the corresponding relationship between real physical information and time. The processing of filtering, amplifying, statistical feature calculation, and correlation analysis of signals in the time domain is collectively referred to as time domain analysis.

When a device fails, its spectrum distribution may change. Like the statistical analysis of time-domain signals, this type of change can be described by statistical analysis of the signal's frequency spectrum.

Given a period of time domain signal $x(t)$, the frequency spectrum of this signal is $y(k), k = 1, \cdots, k$, f_k is the k-th line of the spectrum. Then the time-domain statistical characteristics and frequency-domain statistical features of $x(t)$ are shown in Table 1 [28].

Table 1. The time/frequency-domain statistical features of $x(t)$.

No.	Time-Domain	No.	Frequency-Domain
1	peak : $X_p = \max\{x(n)\}$	14	$F_1 = \frac{1}{K}\sum_{k=1}^{K} y(k)$
2	peak-to-peak : $X_{pp} = \max\{x(n)\} - \min\{x(n)\}$	15	$F_2 = \frac{1}{K-1}\sum_{k=1}^{K}(y(k) - F_1)^2$
3	mean : $\mu = \frac{1}{N}\sum_{n=1}^{N} x(n)$	16	$F_3 = \frac{1}{K(\sqrt{F_2})^3}\sum_{k=1}^{K}(y(k) - F_1)^3$
4	absolute mean : $X_{am} = \frac{1}{N}\sum_{n=1}^{N-1} |x_i|$	17	$F_4 = \frac{1}{K(F_2)^2}\sum_{k=1}^{K}(y(k) - F_1)^4$
5	root amplitude : $X_{ra} = \left(\frac{1}{N}\sum_{n=1}^{N}\sqrt{|x(n)|}\right)^2$	18	$F_5 = \frac{\sum_{k=1}^{K} y(k)f_k}{\sum_{k=1}^{K} y(k)}$
6	standard deviation : $\sigma = \sqrt{\frac{1}{N-1}\sum_{n=1}^{N}[x(n) - \mu]^2}$	19	$F_6 = \sqrt{\frac{\sum_{k=1}^{K} y(k)(f_k - F_5)^2}{K}}$
7	root mean square : $X_{rms} = \sqrt{\frac{1}{N}\sum_{n=0}^{N} x^2(n)}$	20	$F_7 = \sqrt{\frac{\sum_{k=1}^{K} f_k^2 y(k)}{\sum_{k=1}^{K} y(k)}}$
8	skewness : $X_{ske} = \left(\frac{1}{N}\sum_{n=1}^{N}(x(n) - \mu)^3\right)/\sigma^3$	21	$F_8 = \sqrt{\frac{\sum_{k=1}^{K} f_k^4 y(k)}{\sum_{k=1}^{K} f_k^2 y(k)}}$
9	kurtosis : $X_{kur} = \left(\frac{1}{N}\sum_{n=1}^{N}(x(n) - \mu)^4\right)/\sigma^4$	22	$F_9 = \frac{\sum_{k=1}^{K} f_k^2 y(k)}{\sqrt{\sum_{k=1}^{K} y(k)\sum_{k=1}^{K} f_k^4 y(k)}}$
10	peak index : $X_{pi} = X_p/X_{rms}$	23	$F_{10} = \frac{F_6}{F_5}$
11	impulse factor : $X_{imp} = X_p/X_{am}$	24	$F_{11} = \frac{\sum_{k=1}^{K}(f_k - F_5)^3 y(k)}{K(F_6)^3}$
12	margin index : $X_{mi} = X_p/X_{ra}$	25	$F_{12} = \frac{\sum_{k=1}^{K}(f_k - F_5)^4 y(k)}{K(F_6)^4}$
13	waveform index : $X_{wi} = X_{rms}/X_{am}$		

3.1.2. Feature Evaluation and Selection

In this paper, two commonly used feature evaluation methods, Laplacian Score (LS) [29] and Relief-F Score (RFS) [30], are used to evaluate the effectiveness of the time-domain and frequency-domain features of the satellite momentum wheel telemetry signal.

Feature selection is based on two different feature evaluation results, and the feature with the higher evaluation score is taken as the effective feature in the time/frequency domain.

(1) Laplacian Score (LS).

In practical problems, data of the same type are generally close to each other. Under this premise, the importance of describing features can be transformed into evaluating the local retention of features. The Laplace score is based on this idea. Let the data set be $X \in \mathbb{R}^{m \times n}$, L_r is the LS of the r-th feature, f_{ri} is the the r-th feature of the i-th sample. L_r can be calculated as follows.

Step 1: Construct a neighbor graph G containing n nodes, the i-th node corresponds to the i-th sample x_i, if x_i and x_j are close to each other, that is, x_i is within the k-neighbor range of x_j, then an edge is constructed between nodes x_i and x_j. When the data labels are known, edges can be constructed directly between samples of the same type.

Step 2: If nodes x_i and x_j are connected, put $S_{ij} = e^{\frac{\|x_i - x_j\|^2}{t}}$, where t is a suitable constant. Otherwise, put $S_{ij} = 0$. The weight matrix S of the graph models the local structure of the data space.

Step 3: For the r-th feature, the f_r and D can be defined as $f_r = [f_{r1}, f_{r2}, \cdots, f_{rm}]^T$, $D = \text{diag}(S1)$. The matrix $L = D - S$ is often called graph Laplacian. Let

$$\tilde{f}_r = f_r - \frac{f_r^T D 1}{1^T D 1} 1 \tag{5}$$

where $1 = [1, \cdots, 1]^T$.

Step 4: Compute the LS of the r-th feature as follows:

$$L_r = \frac{\tilde{f}_r^T L \tilde{f}_r}{\tilde{f}_r^T D \tilde{f}_r} \tag{6}$$

(2) Relief-F Score (RFS).

The Relief-F Score method is a multi-class variant of the Relief method. The Relief method designs a correlation statistic to measure the importance of features. The statistic is a vector, each component of which corresponds to an initial feature, and the importance of the feature subset is determined by the sum of the relevant statistic components corresponding to each feature in the subset. For each x_i in the data set $X \in \mathbb{R}^{m \times n}$, first find its nearest neighbor $x_{i,nh}$ in the same sample of x_i, which is called guessing nearest neighbor, and then find its nearest neighbor $x_{i,nm}$ from different type samples of x_i, which is called guessing wrong neighbor. The component of the correlation statistic corresponding to the feature is:

$$\delta^{(r)} = \sum_i \left(-\text{diff}\left(x_i^{(r)}, x_{i,nh}^{(r)}\right)^2 + \text{diff}\left(x_i^{(r)}, x_{i,nm}^{(r)}\right)^2 \right) \tag{7}$$

where $x_i^{(r)}$ is the value of the r-th feature of x_i. For x_a and x_b, $\text{diff}\left(x_a^{(r)}, x_b^{(r)}\right)$ depends on the type of the r-th feature. If the r-th feature r is discrete, when $x_a^{(r)} = x_b^{(r)}$, $\text{diff}\left(x_a^{(r)}, x_b^{(r)}\right) = 0$, otherwise $\text{diff}\left(x_a^{(r)}, x_b^{(r)}\right) = 1$. If the r-th feature r is continuous, then $\text{diff}\left(x_a^{(r)}, x_b^{(r)}\right) = \left| x_a^{(r)} - x_b^{(r)} \right|$.

Relief is designed for two classification problems, while Relief-F can handle multiple classification problems. For the sample x_i, if it belongs to the k-th class, the Relief-F method first finds its nearest neighbor $x_{i,nh}$ in the k-th class sample, and then finds a nearest neighbor $x_{i,l,nm}, l \neq k$ of x_i in each class except the k-th class as a guessing wrong neighbor, so the correlation statistic corresponding to the component of the r-th feature is

$$\delta^{(r)} = \sum_i -\text{diff}\left(x_i^{(r)}, x_{i,nh}^{(r)}\right)^2 + \sum_{l \neq k} \left(p_l \times \text{diff}\left(x_i^{(r)}, x_{i,nm}^{(r)}\right)^2 \right) \tag{8}$$

where p_l is the proportion of the l-th class sample in the data set X.

3.2. Complexity Features Based on Huffman-Multi-Scale Entropy (HMSE)

Sample entropy (SampEn) is a new time series complexity characterization parameter proposed by Richman et al. in 2004 [31]. The sample entropy is improved on the basis of approximate entropy, both of which measure the complexity of the time series and the probability of a new pattern generated by the sequence when the dimensionality changes. The greater the probability of generating a new pattern, the more complex the sequence and the higher the entropy value. Compared with other nonlinear dynamic methods such as Lyapunov exponent, information entropy, and correlation dimension, sample entropy has the advantages of short data, strong anti-noise and anti-interference ability, and good consistency within a large range of parameters. Therefore, it has attracted the attention of many scholars and has been frequently used in the field of mechanical signal analysis and fault diagnosis in recent years.

3.2.1. Traditional Multi-Scale Sample Entropy (MSE)

Suppose a time series of length N is $X = \{x_1, x_2, \cdots, x_{N-1}, x_N\}$, and the calculation method of sample entropy is as follows:

Step 1: Construct the time series X into an m-dimensional vector:

$$X(i) = \{x_i, x_{i+1} \cdots, x_{i+m-1}\}, \ i = 1, 2, \cdots, N - m + 1 \qquad (9)$$

Step 2: Define the distance between $X(i)$ and $X(j)$ as $d[X(i), X(j)], (i \neq j)$, which is the largest difference between the two corresponding elements:

$$d[X(i), X(j)] = \max_{k \in (0, m-1)} \left| x(i+k) - x(j+k) \right|, \ (i \neq j) \qquad (10)$$

Step 3: Given a threshold $r > 0$, count the number of $d[X(i), X(j)] < r$ and calculate the ratio to the total number of vectors $N - m$:

$$B_i^m(r) = \frac{1}{N-m} num\{d[X(i), X(j)] < r\} \qquad (11)$$

Step 4: Average all the results obtained by Equation (12):

$$B^m(r) = \frac{1}{N-m+1} \sum_{i=1}^{N-m+1} B_i^m(r) \qquad (12)$$

Step 5: Then m = m + 1, repeat Step1–Step4.
Step 6: Then theoretically the sample entropy of this sequence is:

$$SampEn(m, r) = \lim_{N \to \infty} \left\{ -\ln(\frac{B^{m+1}(r)}{B^m(r)}) \right\} \qquad (13)$$

However, N cannot be infinite in fact, but a finite value. The estimated value of sample entropy is:

$$SampEn(m, r, N) = -\ln(\frac{B^{m+1}(r)}{B^m(r)}) \qquad (14)$$

The sample entropy does not include the comparison of its own data segments, which not only improves the calculation accuracy and saves the calculation time, but also makes the calculation of the sample entropy independent of the data length. In addition, the sample entropy has better consistency. In other words, if one sequence has a higher SampEn than another sequence, then when the parameters m and r are changed, the sequence still has a relatively high SampEn value. However, the disadvantage of sample entropy is that it does not consider the different time scales that may exist in the time series.

To calculate the complexity of the signal at different time scales, Costa et al. proposed multi-scale entropy [32], which aims to extend the sample entropy to multiple time scales to provide additional observation perspectives when the time scale is uncertain. Like other entropy measurement methods, the goal of multi-scale entropy is to evaluate the complexity of time series. One of the main reasons for using multi-scale entropy is that the relevant time scale in the time series is not known. For example, when analyzing a speech signal, it is more effective to count the complexity of the signal under the word time scale than the complexity of the entire speech segment. However, the actual situation is that we often cannot know how many words a certain speech segment contains, or know what time scale should be used to obtain more useful information from the original signal. Therefore, analyzing the problem through multiple time scales will obtain more effective information.

The basic principle of multi-scale entropy (MSE) includes coarse-graining or downsampling the time series, so that the time series can be analyzed at increasingly coarse time resolutions. Given a time series $X = \{x_1, x_2, \cdots, x_{N-1}, x_N\}$ of length N, set the coarse-grained scale to s, then the original time series can be split into i consecutive segments without overlap, where $i = floor(N/s)$, $floor(*)$ means taking the largest integer smaller than $*$. The original sequence can be transformed into a new sequence by calculating the average value of each fragment by Equation (15). Then the MSE of the original sequence can be obtained by solving the sample entropy of the new sequence $Y = \{y_1, y_2, \cdots, y_i\}$ obtained under different s. The process of coarse-graining the time series is shown in Figure 3.

$$y_i = \frac{\sum_{k=1}^{s} x_{(i-1)s+k}}{s} \quad (15)$$

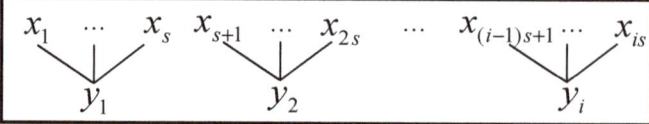

Figure 3. The process of coarse-graining the time series.

3.2.2. The Huffman-Multi-Scale Entropy (HMSE)

According to the process of the calculation of multi-scale sample entropy, the core of this method is to coarse-grain the original time series on different time scales by averaging. Figure 4a shows a satellite momentum wheel voltage telemetry signal with the length of 10,000. This signal is coarse-granulated and averaged on time scales $s = 10$, $s = 50$, and $s = 100$ respectively. The results are shown in Figure 4b–d. As can be seen from Figure 4, the waveform of the new signal obtained by averaging the original signal at different time scales is almost the same.

It can be seen from Figure 4 that the state of the signal changes at about the 4000th sample point in the original signal. However, the use of different coarse-grained scales cannot reflect the difference in signal changes. Therefore, this paper proposes a new improved multi-scale entropy calculation method based on the Huffman mean model. The main innovation of this method is that when the original data is coarse-grained on different time scales, the average value is not taken, but the Huffman average value is taken. This section will introduce the Huffman mean model and the improved multi-scale entropy calculation method based on the Huffman mean model in detail.

(1) Huffman Coding.

In 1952, Huffman proposed an optimum method of coding an ensemble of messages consisting of a finite number of members [33]. A minimum-redundancy code is one constructed in such a way that the average number of coding digits per message is minimized. The process of Huffman coding is as follows.

Step 1: Given a sequence containing n kinds of symbols. Suppose the set of symbol types is $S_0 = \{s_1^0, s_2^0, \cdots, s_i^0, \cdots s_n^0\}, i = 1, 2, \cdots, n$. The probability of each symbol appearing is $P_0 = \{p_1^0, p_2^0, \cdots, p_i^0, \cdots p_n^0\}, i = 1, 2, \cdots, n$, and $\sum_{i=1}^{n} p_i^0 = 1$.

Step 2: Set the iteration parameter to t, the maximum value of t is $n - 1$ and the initial value of t is 0. The symbol sequence and the corresponding probability at the beginning of the k-th iteration are S_{k-1} and P_{k-1}. The symbol sequence and the corresponding probability at the end of the k-th iteration are S_k and P_k.

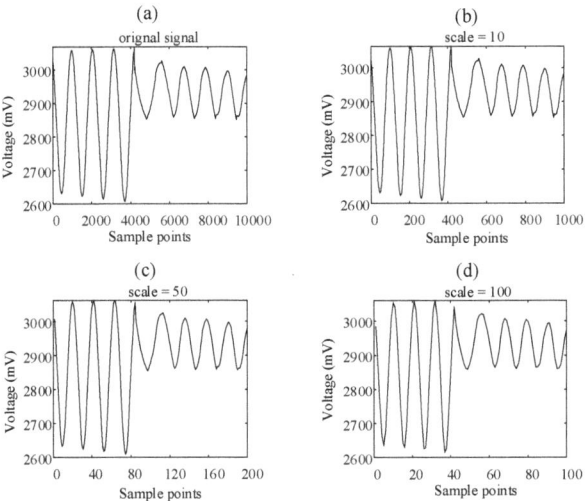

Figure 4. The average value of voltage telemetry under different scales: (**a**) original signal, (**b**) scale = 10, (**c**) scale = 50, (**d**) scale = 100.

Step 3: When $t = k$, arrange the symbol S_{k-1} in ascending order of probability P_{k-1} as $S_{k-1} = \{s_1^{k-1}, s_2^{k-1}, \cdots, s_i^{k-1}, \cdots s_{n-k+1}^{k-1}\}$. Then the probability P_{k-1} is also rearranged accordingly as $P_{k-1} = \{p_1^{k-1}, p_2^{k-1}, \cdots, p_i^{k-1}, \cdots p_{n-k+1}^{k-1}\}$.

Step 4: Take the two symbols s_1^{k-1} and s_2^{k-1} with the least probability in the symbols sequence. Encode the symbol s_2^{k-1} with higher probability into "1" and the symbol s_1^{k-1} with lower probability as "0". Add the probabilities p_1^{k-1} and p_2^{k-1} of the s_1^{k-1} and s_2^{k-1} as the probability p^* of the new symbol s^*.

Step 5: Delete s_1^{k-1} and s_2^{k-1} from S_{k-1}, and add s^* into S_{k-1}. Then the S_{k-1} turns into S_k, and the size of S_k is $n - k$. Delete p_1^{k-1} and p_2^{k-1} from P_{k-1}, and add p^* into P_{k-1}. Then the P_{k-1} turns into P_k, and the size of P_k is also $n - k$.

Step 6: Repeat the Step 3 to Step 5 until $t = n - 1$. Then the symbols sequence will be $S_{n-1} = \{s_1^{n-1}\}$ and the probability will be $P_{n-1} = \{p_1^{n-1}\}, p_1^{n-1} = 1$.

It can be seen from the above coding process that the symbol with the lower probability in the original signal has the longer Huffman code length. Conversely, the symbol with the higher probability has the shorter Huffman code length. The complexity of the probability distribution of the signal can be described by solving the Huffman average code length of the original signal. Based on the above-mentioned Huffman coding process, the method to further calculate the average Huffman coding length is as follows.

Backtrack from the symbol s_1^{n-1} with the probability of $p_1^{n-1} = 1$ to each source symbol and record 0/1 in the backtracking path. The Huffman code of s_i^0 is c_i. The average Huffman coding length L^* can be calculated according to the length of c_i and the corresponding probability p_i^0 as Equation (16). $L(c_i)$ is the length of c_i.

$$L^* = \sum_{i=1}^{n} p_i^0 * L(c_i) \qquad (16)$$

For a set of source symbols $S_0 = \{s_1, s_2, s_3, s_4, s_5, s_6\}$ with probability $P_0 = \{0.35, 0.28, 0.14, 0.13, 0.07, 0.03\}$, the process of Huffman coding is shown in Table 2. The average Huffman coding length of S_0 can be calculated as 2.33 as follows.

$$L^*(S_0) = (0.35 + 0.28 + 0.14) * 2 + 0.13 * 3 + (0.07 + 0.03) * 4 = 2.33$$

Table 2. An example of the Huffman coding process.

S_0	P_0	Huffman Coding					c_i	$L(c_i)$
s_1	0.35	→ 0.35	→ 0.35	→ 0.37	→ 0.63 → 1	1	11	2
s_2	0.28	→ 0.28	→ 0.28	→ 0.35 → 1	→ 0.37 → 0		10	2
s_3	0.14	→ 0.14	→ 0.23 → 1	→ 0.28 → 0			00	2
s_4	0.13	→ 0.13 → 1	→ 0.14 → 0				011	3
s_5	0.07	→ 1 → 0.1 → 0					0101	4
s_6	0.03	→ 0					0100	4

(2) Huffman Mean Model.

The basic principle of Huffman coding and the calculation method for solving the average Huffman coding length were introduced above. In this paper, a new Huffman mean model based on the Huffman coding is proposed for the problem of satellite anomaly detection. For a sequence $T = \{t_1, t_2, \cdots, t_i, \cdots, t_n\}$, the expression of the Huffman mean model is shown in Equation (17).

$$HM(T) = \begin{cases} T' = T/sum(T) \\ C = Huffman\_coding(T') \\ \ell = L(C) \\ Huffman\_mean = sum(T * (\ell/sum(\ell))) \end{cases} \qquad (17)$$

where $HM(T)$ is the Huffman mean value of T, $T' = T/sum(T)$ means to convert the original time series into a probability series, $C = Huffman\_coding(T')$ represents the Huffman coding result of the probability sequence T', $C = \{c_1, c_2, \cdots, c_i, \cdots, c_n\}, i = 1, 2, \cdots, n$, c_i is the Huffman code corresponding to t_i in the original sequence, $\ell = L(C)$ means to calculate the length of each c_i, $Huffman\_mean = sum(T * (\ell/sum(\ell)))$ represents the Huffman mean value of the sequence T considering the length weight of the Huffman code.

(3) The Improved Method of Huffman-multi-scale Entropy.

The Figure 5 shows the calculation process of Huffman-multi-scale entropy. The inputs of both two methods are original signal $X = \{x_1, x_2, \cdots, x_{N-1}, x_N\}$, the scale sequence $Scale = \{s_1, s_2, \cdots, s_p\}$ and the parameter set $\theta = \{m, r\}$, usually $0.1std(X) < r < 0.2std(X)$. Compared with the classic MSE, the Huffman-multi-scale entropy method proposed in this paper adopts the coarse-grained method based on the Huffman mean model.

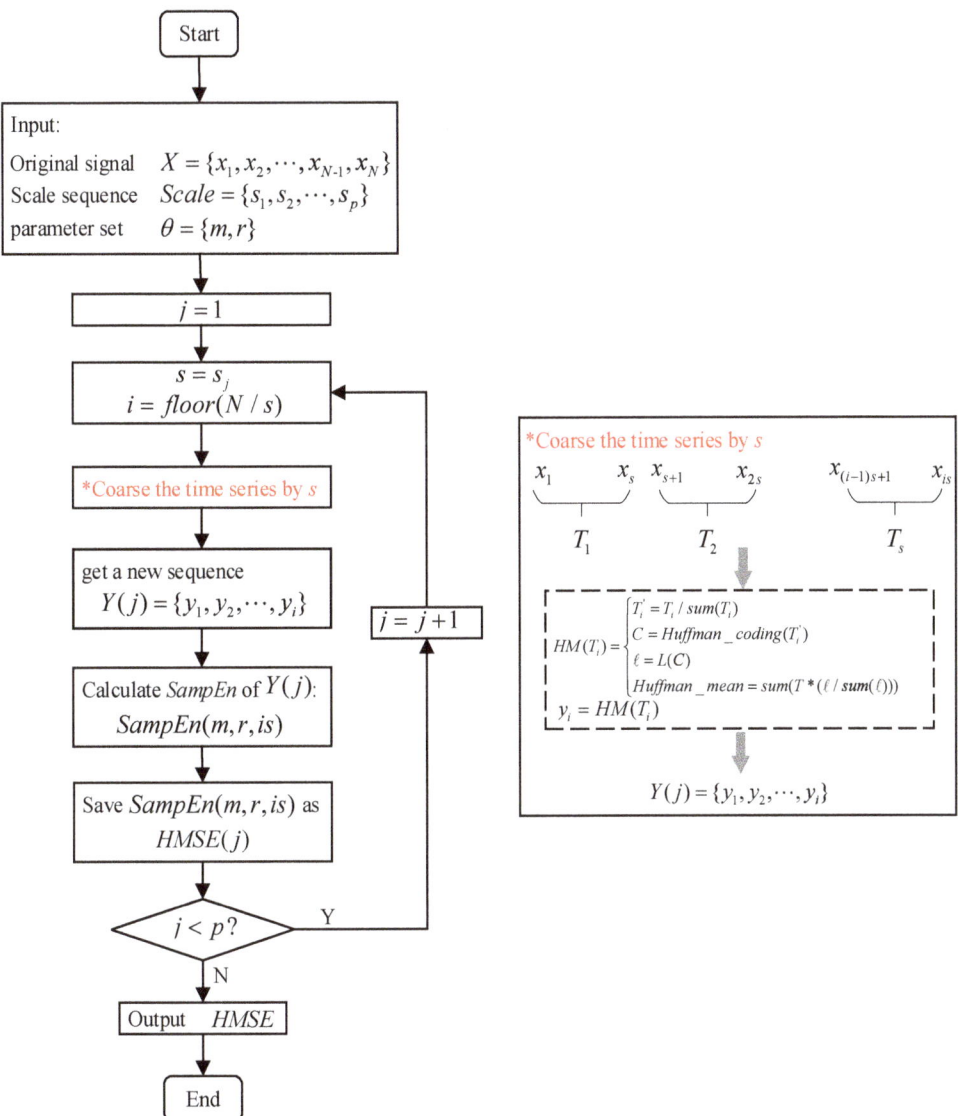

Figure 5. The process of the improved method of Huffman-multi-scale Entropy.

Figure 6 shows the same satellite momentum wheel voltage telemetry signal with the length of 10,000. This signal is coarse-granulated and calculated by Huffman mean model on time scales $s = 10$, $s = 50$, and $s = 100$, respectively. Obviously, the coarse-grained method based on the Huffman mean model can enhance the difference of signal changes at different time scales.

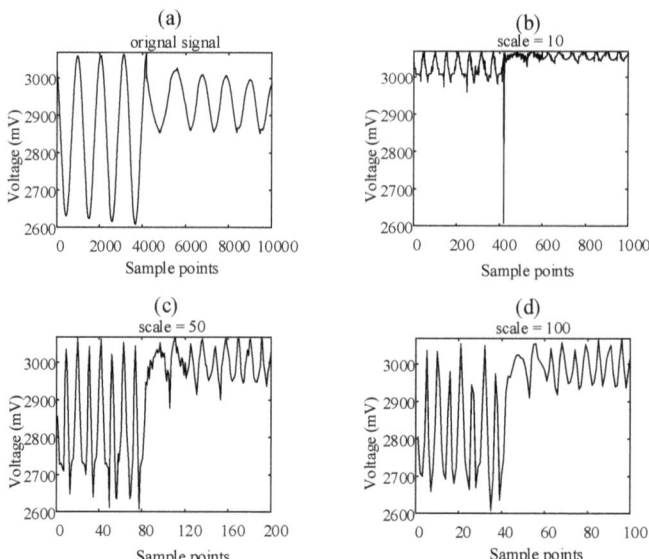

Figure 6. The Huffman average code length of voltage telemetry under different scales: (**a**) original signal, (**b**) scale = 10, (**c**) scale = 50, (**d**) scale = 100.

4. Anomaly Detection Method Based on Multi-Class SVM

4.1. Multi-Class SVM

A support vector machine (SVM) is widely used in classification problems. The basic idea is to find a hyperplane so that all sample points in the positive and negative categories are farthest from the plane, and points that are far enough from the plane can basically be correctly classified. Therefore, if the points closer to the hyperplane are as far away as possible from the hyperplane, a better classification effect can be achieved.

This article uses the most interval classifier to achieve two-class SVM, and then uses the directed acyclic graph (DAG) method to achieve the multi-class SVM based on two-class SVM.

Set the dataset as $\{(x_i, y_i) | i = 1, 2, \cdots, N\}, x_i \in R_n, y \in \{-1, +1\}$, the hyperplane is $w^T x + b = 0$, Then the distance from the support vector to the hyperplane is $w^T x + b = y$, which can be written as

$$\frac{|y(w^T x + b)|}{||w||_2} = \frac{1}{||w||_2} \tag{18}$$

The SVM model keeps all the points on both sides of the support vector of their respective categories, while keeping away from this hyperplane. It can be seen from Equation (18) that when $||w||_2$ is the smallest, the interval is the largest. Introduce the penalty parameter λ for misclassification and the relaxation factor ξ that allows misclassification, and the objective function can be

$$\min \tfrac{1}{2}||w||_2^2 + \lambda \sum_{i=1}^{N} \xi_i$$
$$s.t. \quad y^i(w^T x + b) \geq 1 - \xi_i, i = 1, 2, \cdots, N \tag{19}$$
$$\xi_i \geq 0, i = 1, 2, \cdots, N$$

According to Lagrange's duality, the optimization objective can be converted into an equivalent dual problem. The Equation (19) can be transformed into:

$$\min_{\alpha} \frac{1}{2} \sum_{i,j=1}^{N} y_i y_j \alpha_i \alpha_j K\langle x_i, x\rangle - \sum_{j=1}^{N} \alpha_j$$
$$s.t. \quad y^i(w^T x + b) \geq 1 - \xi_i, i = 1, 2, \cdots, N$$
$$\xi_i \geq 0, i = 1, 2, \cdots, N \quad (20)$$

where $K\langle x_i, x\rangle$ is the kernel function. The radial basis function is used as the kernel function in this paper, $K\langle x_i, x\rangle = \exp\left\{-\frac{\|x_i - x\|^2}{2\sigma^2}\right\}$, σ is the kernel function parameter. Then the decision function is

$$f(x) = w^T x + b = \text{sgn}[\sum_{i=1}^{N} y_i \alpha_i K\langle x_i, x\rangle + b], 0 < \alpha_i < \lambda \quad (21)$$

Among the multi-class SVM methods, one is the direct solution method, but this method has high time complexity and is difficult to implement. It is not suitable for a large amount of data. The other one is to combine multiple two-class SVM models into a multi-class SVM model. In this paper, a directed acyclic graph method is used to construct a multi-class SVM.

The DAG method uses the "competition" rule. For n types, the height of the decision tree is $n - 1$. Put the classes that are easy to distinguish on the upper layer, and the classes that are difficult to distinguish on the lower layer. The schematic diagram of DAG method for five-class SVM is shown in Figure 7.

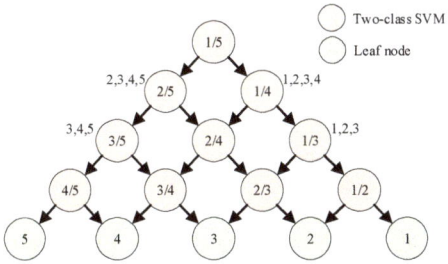

Figure 7. The schematic diagram of DAG method for five-class SVM model.

4.2. Improved Adaptive Particle Swarm Optimization (APSO)

Kennedy and Eberhart first proposed particle swarm optimization (PSO) in 1995 [34]. PSO is an algorithm for finding the optimal solution inspired by the foraging behavior of bird groups. In the PSO algorithm, each particle represents a feasible solution of a function to be optimized, and the movement of the particle is restricted by two aspects: speed and position. The speed constrains the distance of particle movement, while the position constrains the direction of particle movement. Each particle's movement is given a fitness function to evaluate the particle's location. Under the control of constraint conditions and evaluation function, the particles search for a better area in the process of moving. After many iterations, they gather near the optimal solution. The particle velocity and position update formula are as follows:

$$v_{id}^{k+1} = \omega v_{id}^k + c_1 r_1 (p_{id} - x_{id}^k) + c_2 r_2 (p_{gd} - x_{id}^k) \quad (22)$$

$$x_{id}^{k+1} = x_{id}^k + v_{id}^{k+1} \quad (23)$$

where v_{id}^k is the current velocity of the d-th component in the i-th particle, v_{id}^{k+1} is the next velocity of the d-th component in the i-th particle, ω is the inertia weight, $\omega \geq 0$, c_1 and c_2 are the acceleration constant of the particle, r_1 and r_2 are random numbers between 0 and 1, $r_1, r_2 = random(0,1)$, p_{id} represents the best position of the d-th component of the i-th particle, p_{gd} represents the best position of the d-th component of all particles, x_{id}^k is the current position of the d-th component in the i-th particle, and x_{id}^{k+1} is the next position of the d-th component in the i-th particle.

PSO has the advantages of fewer parameters and fast convergence, but it also has shortcomings such as premature convergence and falling into local optimum. It can be seen from Equations (22) and (23) that the inertia weight ω determines the relationship between the next flight distance and the current flight distance, which further affects the position after the flight. The larger the ω, the stronger the particle's flying ability in the solution space, which is conducive to searching in the global scope. The smaller the ω, the smaller the flight length, and the stronger the search ability of the particles in a local area, which is conducive to the convergence of the algorithm. However, if the value of ω is too large, it will easily cause the algorithm to skip the optimal solution or oscillate near the optimal solution, which will lead to the premature convergence; if ω is too small, the algorithm will easily fall into a local optimum.

The inertia weight ω should be a larger value at the beginning of the iteration to ensure a strong global search ability and the ability to jump out of the local optimum. However, in the later stage of the algorithm iteration, smaller ω should be used to ensure strong local search capabilities, which is conducive to the convergence of the algorithm.

In response to the above problem, this paper proposes a strategy for adaptively changing the ω according to the number of iterations and the current fitness value. The formula is as follows:

$$\omega_{id}^{k+1} = \begin{cases} \omega_0 - e^{-k/K} * \frac{f_{\max}^k - f_{id}^k}{f_{\max}^k - f_{avg}^k}, & f_{id}^k \leq f_{avg}^k \\ \omega_0 + e^{-k/K} * \frac{f_{id}^k - f_{\min}^k}{f_{avg}^k - f_{\min}^k}, & f_{id}^k > f_{avg}^k \end{cases} \quad (24)$$

where k is the current number of iterations, $k+1$ is the next number of iterations, K is the maximum number of iterations, ω_{id}^{k+1} is the inertia weight for the next iteration for the d-th component of the i-th particle, ω_0 is the initial value of ω, $\omega_0 = 0.5$ in this paper, f_{id}^k is the fitness value of the d-th component of the i-th particle obtained in the k-th iteration, $f_{\max}^k$ is the maximum fitness value in the k-th iteration, $f_{\min}^k$ is the minimum fitness value in the k-th iteration, f_{avg}^k is the average fitness value in the k-th iteration.

At the beginning of the iteration, the weight of the particle changes greatly, and as the number of iterations of the particle increases, the weight change decreases. At the same time, the weight change is determined by the fitness function. When the particle fitness is less than or equal to the average fitness, that is, when the accuracy of the classification model is greater than or equal to the average accuracy, the inertia weight decreases; when the particle fitness is greater than the average fitness, the accuracy of the classification model is lower than the average accuracy, and the inertia weight increases.

The increase or decrease of the inertia weight is determined by the number of iterations. At the beginning of the iteration, the increase or decrease of the weight is large, which is convenient for searching in the global and optimal solution neighborhood. At the later stage of the iteration, the increase or decrease of the weight is small. The increase of the weight can avoid falling into the local optimal solution for random search, and the decrease of the weight facilitates the local fine search.

4.3. The Algorithm of the Proposed APSO-SVM

In this paper, the improved Adaptive Particle Swarm Optimization (APSO) is used to optimize the penalty factor λ and the kernel function parameter σ in the SVM. The specific steps of APSO-SVM are as follows.

Step 1: Input the dataset with labels.
Step 2: Divide the dataset into training set and test set, then normalize both two sets.
Step 3: Population initialization. Set the number of particles in the initial population as n. Set the range of penalty factor λ to $[\lambda_{\min}, \lambda_{\max}]$. Set the range of kernel function parameter σ to $[\sigma_{\min}, \sigma_{\max}]$. Initialize the parameter set $\theta = \{\omega_0, c_1, c_2, K\}$. Initialize the position x_i^0, the speed v_i^0, the optimal position p_{id} of i-th particle and the global optimal position p_{gd}. Set fitness error ε.
Step 4: Calculate the corresponding inertia weight ω_{id}^{k+1} according to Equation (24) in the adaptive adjustment strategy. Update the velocity v_{id}^{k+1} and position x_{id}^{k+1} of the particles according to Equations (22) and (23). Determine whether λ and σ are in $[\lambda_{\min}, \lambda_{\max}]$ and $[\sigma_{\min}, \sigma_{\max}]$ respectively. If $\lambda < \lambda_{\min}$, set $\lambda = \lambda_{\min}$. If $\lambda > \lambda_{\max}$, set $\lambda = \lambda_{\max}$. If $\sigma < \sigma_{\min}$, set $\sigma = \sigma_{\min}$. If $\sigma > \sigma_{\max}$, set $\sigma = \sigma_{\max}$.
Step 5: If $f_i > f(p_i)$ or $|f_i - f(p_i)| \leq \varepsilon, \lambda(x_i) < \lambda(p_i)$, update p_i. If $f_i > f(p_g)$ or $|f_i - f(p_g)| \leq \varepsilon, \lambda(x_i) < \lambda(p_g)$, update p_g. The expression of $f(*)$ is shown in Equation (21).
Step 6: If $k < K$, repeat the step 4 to step 6. If $k \geq K$, end the APSO.
Step 7: Use the optimal solution (λ^*, σ^*) to create the SVMs model and use this model for classification.
The flow chart of APSO-SVM is shown in Figure 8.

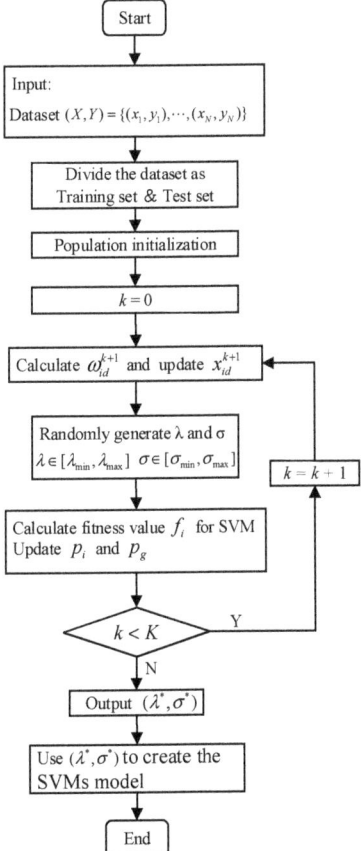

Figure 8. Flow chart of APSO-SVM.

5. Case Study of Anomaly Detection

5.1. Data Description

The data set used in this article is from a satellite's telemetry voltage value of its momentum wheel. In this data set, five types of sample with different health status are screened out. Stable Change (large) indicates that the momentum wheel voltage value continuously and steadily changes with a large change amplitude. Stable Change (small) indicates that the momentum wheel voltage value continuously and steadily changes with a small change amplitude. Large to Small indicates that the amplitude of the momentum wheel voltage change smoothly transitions from large to small. The above three types of sample all represent that the momentum wheel is in a normal state. Irregular Change indicates that the voltage of the momentum wheel changes irregularly. Sudden Change indicates that the voltage of the momentum wheel has a sudden change, such as the voltage suddenly jumping to 0. Irregular Change and Sudden Change represent that the momentum wheel is in an abnormal state. The time-domain waveforms of different types of data are shown in the Figure 9.

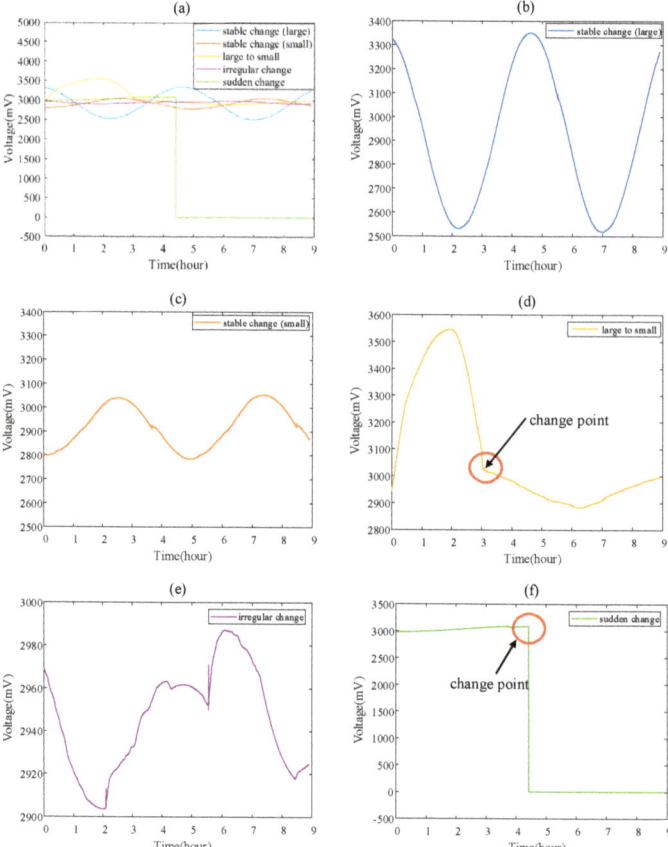

Figure 9. The waveform of five different types of voltage telemetry data for the momentum wheel: (**a**) comparison of 5 types of data in time-domain, (**b**) the waveform of Stable Change (large), (**c**) the waveform of Stable Change (small), (**d**) the waveform of Large to Small, (**e**) the waveform of Irreg-ular Change, (**f**) the waveform of Sudden Change.

To verify the effectiveness of the method proposed in this article, the training set and test set used in this article are shown in the Table 3.

Table 3. Label description of the momentum wheel voltage telemetry dataset.

Class	Label	Health Status	Training Set	Test Set
1	Stable Change (large)	Normal	400	200
2	Stable Change (small)	Normal	400	200
3	Large to Small	Normal	400	200
4	Irregular Change	Abnormal	400	200
5	Sudden Change	Abnormal	400	200

5.2. Feature Extraction and Selection

5.2.1. Time/Frequency Domain Feature Extraction and Selection

According to the time-domain feature and frequency-domain feature calculation methods shown in Table 1, the time-frequency feature values of the five types of momentum wheel voltage telemetry data are calculated. The time-domain features are shown in Figure 10, and the frequency-domain features are shown in Figure 11.

According to the time-frequency feature statistical feature distribution diagrams of different types of data in Figures 10 and 11, the time-domain feature distribution of SC is very scattered, but the frequency-domain feature distribution is relatively more concentrated. Intuitively, peak and peak-to-peak in the time domain feature can distinguish five types of data to a certain extent, and F3 and F4 in the frequency domain feature can also distinguish five types of sample to a certain extent.

In order to quantify the ability of different feature values to distinguish samples, we use the feature evaluation method (Laplacian Score and Relief-F Score) in Section 3.1.2 to score the above 25 types of time-domain features and frequency-domain features. The evaluation results are shown in Figure 12. Comprehensively considering the evaluation results of LS and RFS, this paper chooses nine features (peak, peak-to-peak, skewness, kurtosis, F3, F4, F8, F10 and F11), which have higher scores in two evaluation methods, as part of the feature sequence. These high-scoring features describe the amplitude characteristics, fluctuation characteristics and spectral density characteristics of the voltage telemetry signals.

5.2.2. Complexity Feature Analysis

To verify the effectiveness of the proposed complexity feature extraction method of Huffman-multi-scale entropy, this paper analyzes the sample entropy under different sample lengths and different scales.

Taking a normal type data Stable Change (large) as an example to study the impact of sample length on complexity characteristics, the sample length is taken from 5000 to 25,000 at intervals of 2000. Figure 13 shows the results of calculating the multi-scale entropy and Huffman-multi-scale entropy with each sample length respectively. The scale is from 10 to 300 at intervals of 10. It can be found that when the sample length is 9000 to 15,000, both methods can achieve higher sample entropy. Therefore, this paper selects the sample length as 10,000.

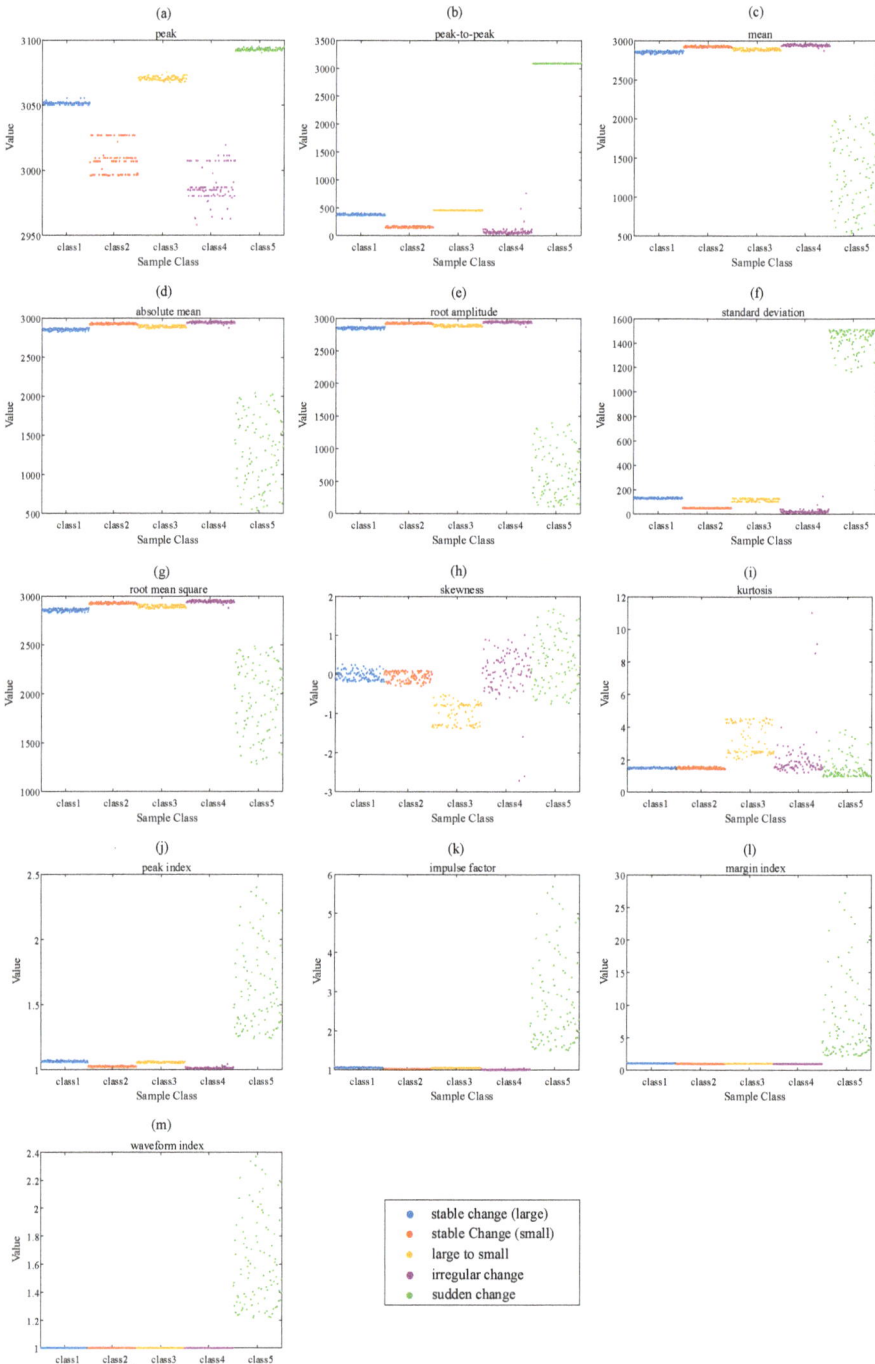

Figure 10. The time-domain features of the five types of momentum wheel voltage telemetry data: (**a**) peak, (**b**) peak-to-peak, (**c**) mean, (**d**) absolute mean, (**e**) root amplitude, (**f**) standard deviation, (**g**) root mean square, (**h**) skewness, (**i**) kur-tosis, (**j**) peak index, (**k**) impulse factor, (**l**) margin index, (**m**) waveform index.

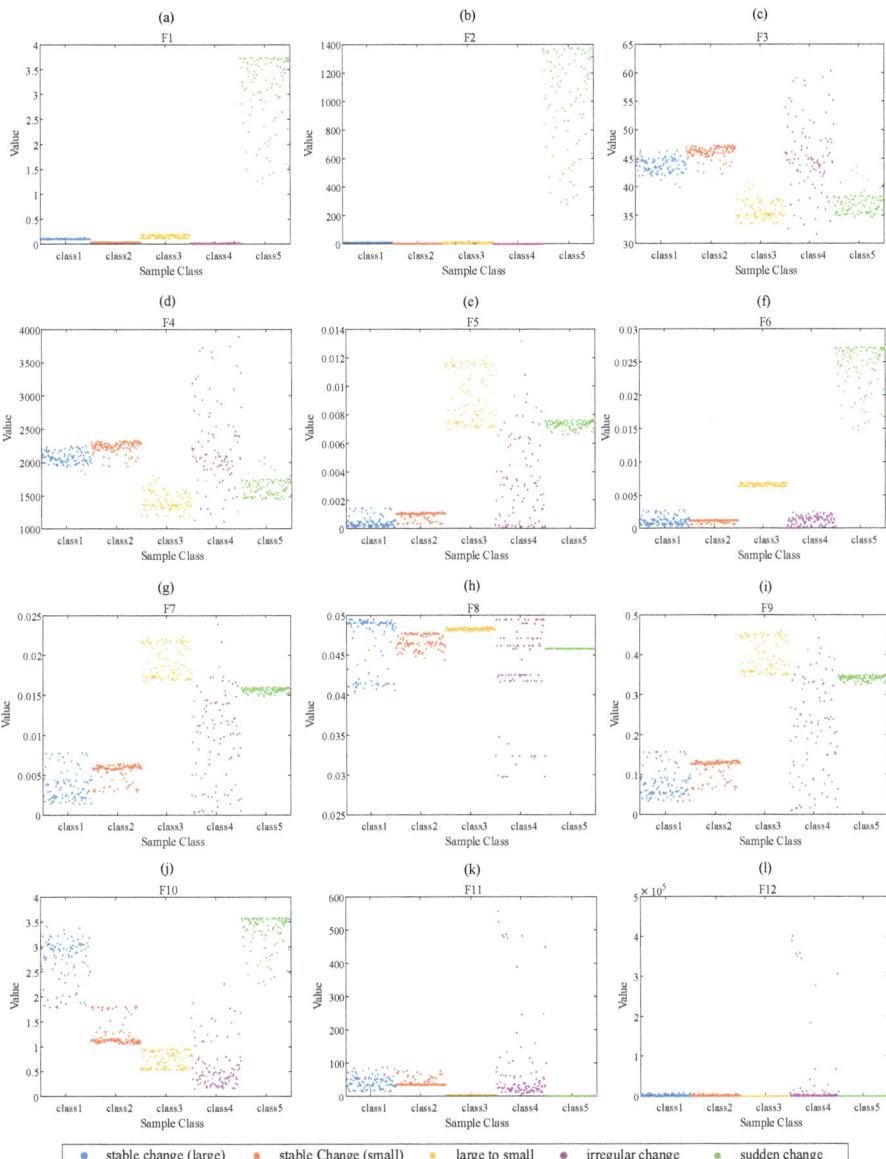

Figure 11. The frequency-domain features of the five types of momentum wheel voltage telemetry data: (**a**) F1, (**b**) F2, (**c**) F3, (**d**) F4, (**e**) F5, (**f**) F6, (**g**) F7, (**h**) F8, (**i**) F9, (**j**) F10, (**k**) F11, (**l**) F12.

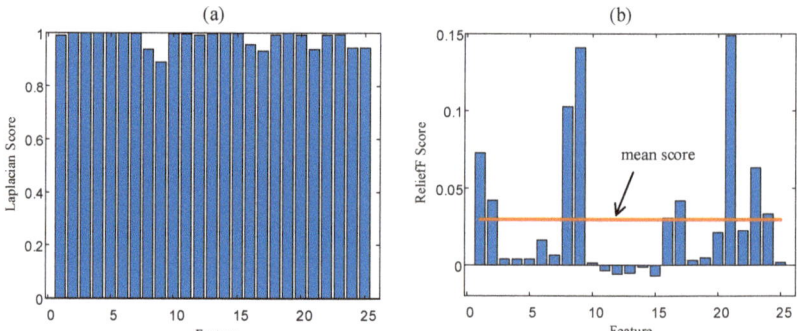

Figure 12. The time-domain (**a**) and frequency-domain (**b**) features evaluation results.

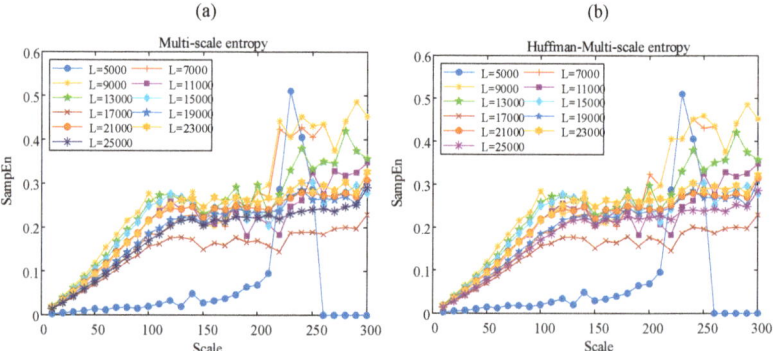

Figure 13. The results of the multi-scale entropy (**a**) and Huffman-multi-scale entropy (**b**) with different sample length.

At the same time, this paper also calculates the multi-scale entropy and Huffman-multi-scale entropy of different types of momentum wheel voltage telemetry signals when the sample length is 10,000. The scale ranges from 10 to 300, with an interval of 10. The calculation result is shown in Figure 14. It can be seen from Figure 14 that, for normal data, the results of multi-scale entropy and Huffman-multi-scale entropy are close. For Irregular Change data, the value of Huffman-multi-scale entropy is significantly lower than that of multi-scale entropy. This shows that Huffman-multi-scale entropy has better distinguishing ability for data with high complexity. It is worth noting that, for the abnormal data of Sudden Change type, the data itself has pulse characteristics, which causes the fluctuation characteristics of the data before and after the sudden change to be concealed to a certain extent. Therefore, both multi-scale entropy and Huffman-multi-scale entropy can well describe the characteristics of increased complexity caused by sudden and large changes in data.

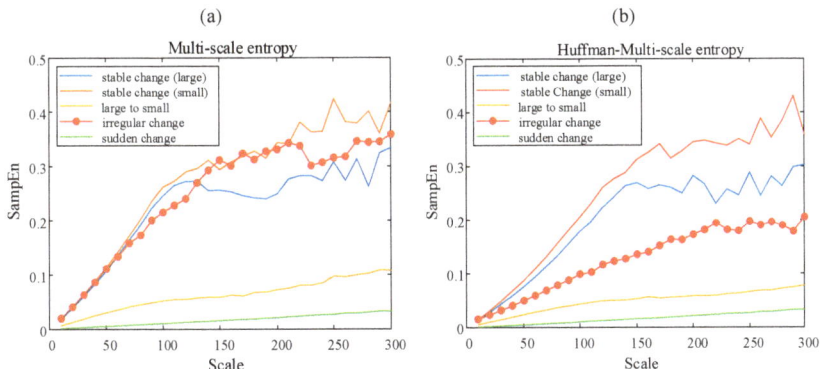

Figure 14. Results of the multi-scale entropy (**a**) and Huffman-multi-scale entropy (**b**) for different data type.

Based on the above analysis, this paper selects the sample length as 10,000. The feature sequence (HMSE + T/F) is composed of 30 complexity features (scale from 10 to 300 at intervals of 10) and nine time/frequency-domain features.

5.3. Anomaly Detection Results and Discussion

This paper uses a five-class SVM model based on the DAG method, and uses the proposed APSO to train the classification model. To verify the effectiveness of the proposed method on the spacecraft anomaly detection problem, this paper not only calculates the recognition accuracy (RA) of the classification model for each category, but also calculates the detection rate (DR), false alarm rate (FAR) and the missed alarm rate (MAR). The calculation method of RA, DR, FAR and MAR are shown in Equations (25)–(28).

$$RA = \frac{Num(predicted = true)}{Num(true)} * 100\% \quad (25)$$

$$DR = \frac{Num(NN + FF)}{Num(true)} * 100\% \quad (26)$$

$$FAR = \frac{Num(NF)}{Num(N)} * 100\% \quad (27)$$

$$MAR = \frac{Num(FN)}{Num(F)} * 100\% \quad (28)$$

where $Num(predicted = true)$ is the total number of category predictions that are exactly the same as the true value, $Num(true)$ is the total number of the test samples, $Num(NN + FF)$ is the total number of the real normal data predicted as normal data and the real abnormal data predicted as abnormal data, $Num(NF)$ is the total number of real normal data predicted as abnormal data, $Num(N)$ is the total number of real normal data, $Num(FN)$ is the total number of real abnormal data predicted as normal data, and $Num(F)$ is the total number of real abnormal data.

Figure 15 shows the corresponding part of the false alarm rate and the missed alarm rate in the confusion matrix. C(large) is the Stable Change (large), C(small) is the Stable Change (small), L-S is the Large to Small, IC is the Irregular Change, and SC is the Sudden Change.

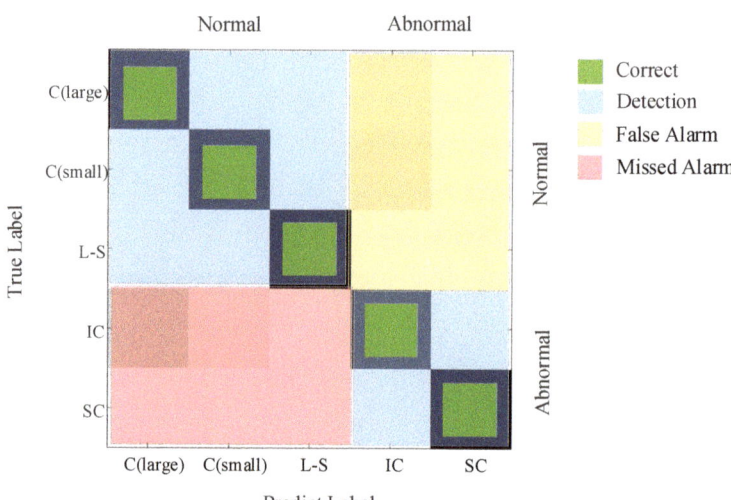

Figure 15. Correct, false alarms and missed alarms in the confusion matrix.

This paper calculates the confusion matrix of the abnormal detection of the momentum wheel voltage telemetry signal calculated by HMSE-T/F-APSO-SVM, MSE-T/F-APSO-SVM and MSE-T/F-PSO-SVM. The results are shown in Figure 16. It can be seen from the Figure 16, the identification accuracy of Sudden Change by the above three methods can all reach 100%. This result is consistent with the conclusion of the qualitative analysis of eigenvalues in the previous article. The probability of MSE-T/F-PSO-SVM identifying Stable Change (large) and Stable Change (small) as Irregular Change reaches 10.33% and 11.33%, respectively. At the same time, the probability of MSE-T/F-PSO-SVM identifying Irregular Change as Stable Change (large) and Stable Change (small) reaches 16.67% and 7.67%, respectively. The probability of HMSE-T/F-APSO-SVM and MSE-T/F-PSO-SVM identifying Stable Change (large) and Stable Change (small) as Irregular Change are 0%. At the same time, the probability of MSE-T/F-PSO-SVM identifying Irregular Change as Stable Change (large) and Stable Change (small) reaches 16.67% and 7.67%, respectively. The distinction between Stable Change (large) Stable Change (small) and Irregular Change can be effectively improved by calculating Huffman-multi-scale entropy. This conclusion is also consistent with the result in Figure 14.

In addition, the recognition accuracy of Large to Small and Sudden Change of the three methods has reached 100%, which shows that the feature sequence and anomaly detection model selected in this paper have strong sensitivity to signals with a definite change rule.

To further verify that the method proposed in this paper can effectively improve the accuracy of spacecraft anomaly detection and reduce the rate of false alarms and missed alarms, this paper compares the proposed method with other methods, and calculates the anomaly detection under different processing methods. Principal Component Analysis (PCA), Random forest (RF), Logistic Regression (LR), K-Nearest Neighbor (KNN) and Multilayer perceptron (MLP) are used in this paper. The results of the recognition accuracy, false alarm rate and missed alarm rate of different methods are shown in Table 4.

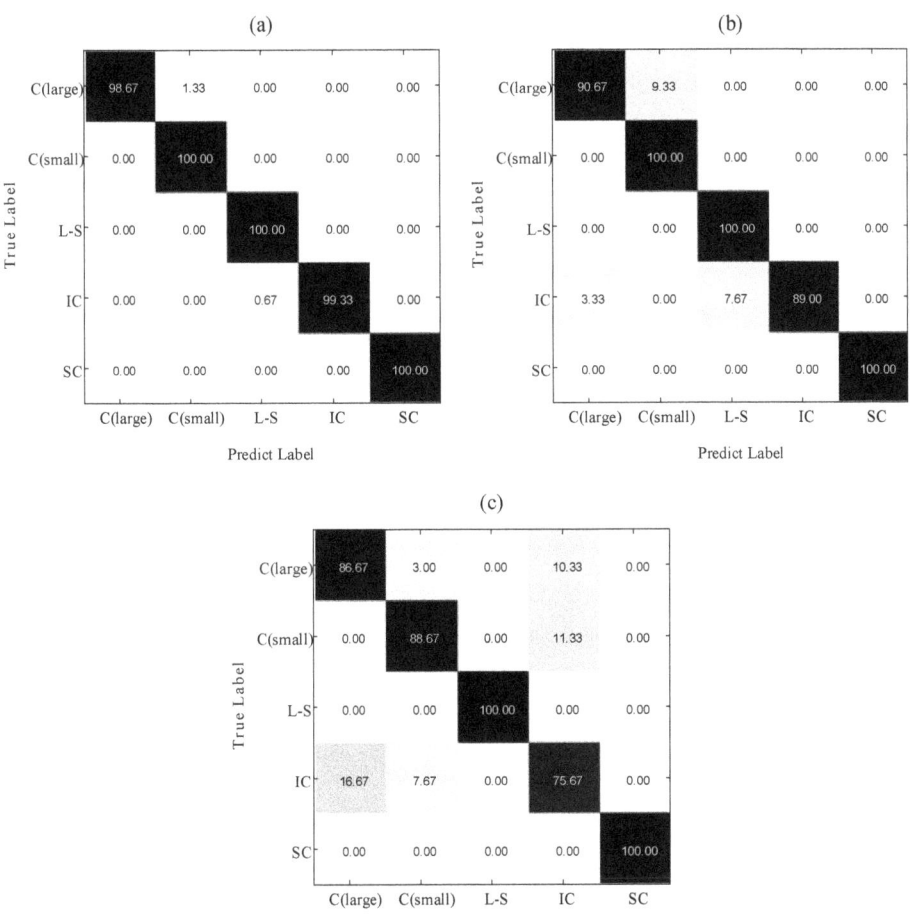

Figure 16. The confusion matrix of the abnormal detection for the momentum wheel voltage telemetry signal calculated by different processing methods (%): (**a**) HMSE-T/F-APSO-SVM, (**b**) HMSE-T/F-PSO-SVM, (**c**) MSE-T/F-PSO-SVM.

Table 4. Results of the recognition accuracy, detection rate, false alarm rate and missed alarm rate of different methods (%).

Methods	Recognition Accuracy	Detection Rate	False Alarms Rate	Missed Alarms Rate
HMSE + T/F + APSO + SVM	99.60 ± 0.28	99.87 ± 0.32	0.00 ± 0.00	0.34 ± 0.01
HMSE + T/F + PSO + SVM	95.93 ± 0.36	97.80 ± 0.47	0.00 ± 0.00	5.50 ± 0.12
MSE + T/F + PSO + SVM	90.20 ± 0.74	90.80 ± 0.39	7.22 ± 0.85	12.17 ± 1.68
Original data + SVM	80.55 ± 2.01	83.61 ± 2.34	15.88 ± 1.58	17.16 ± 2.54
Original data + PCA	75.26 ± 4.97	77.31 ± 6.28	23.32 ± 2.69	21.74 ± 4.71
HMSE + T/F + PCA	81.75 ± 2.96	85.36 ± 3.12	12.61 ± 1.47	17.68 ± 3.63
Original data + RF	76.95 ± 4.38	80.73 ± 5.26	19.54 ± 1.85	18.86 ± 1.95
HMSE + T/F+ RF	84.28 ± 1.98	87.62 ± 2.41	14.46 ± 1.74	9.26 ± 0.36
Original data + LR	70.36 ± 4.33	72.95 ± 8.57	20.35 ± 3.54	37.1 ± 5.43
HMSE + T/F+ LR	83.78 ± 4.05	85.74 ± 4.69	10.68 ± 0.79	19.63 ± 1.37
Original data + KNN	70.37 ± 4.69	72.83 ± 5.78	24.79 ± 4.75	30.74 ± 6.35
HMSE + T/F+ KNN	83.55 ± 1.38	86.28 ± 1.54	14.82 ± 1.24	12.07 ± 1.73
Original data + MLP	86.69 ± 2.17	87.94 ± 2.24	10.67 ± 2.42	14.14 ± 1.58
HMSE + T/F + MLP	95.25 ± 0.57	96.63 ± 0.68	0.00 ± 0.00	8.42 ± 0.83

From the results in Table 4, it can be seen that: First, the recognition accuracy and detection rate of the method proposed in this paper can reach 99.60% and 99.87%, which are higher than other methods listed in the table, and the false alarm rate is reduced to 0, while the false alarm rate is reduced to 0.34%, which are lower than other methods. Second, the detection method based on the feature sequence (HMSE + T/F) has a higher recognition accuracy and detection rate as well as lower false alarm rate and missed detection rate than the detection method based on the original data. Third, by comparing the standard deviation of various indicators, it can be found that the feature sequence based on Huffman-multi-scale entropy and time-frequency domain features proposed in this paper can effectively improve the stability of the detection method.

6. Conclusions

In this research, we propose a new detection framework for anomaly detection based on spacecraft telemetry data. Due to the very low frequency characteristics of telemetry data, most frequency analysis methods are not suitable for spacecraft anomaly detection. Therefore, this paper first proposes a feature sequence construction method based on time-domain and frequency-domain feature screening and complexity feature fusion. On this basis, a new method of Huffman-multi-scale entropy (HMSE) based on the Huffman coding principle is proposed. To improve the classification accuracy of SVM, this paper adopts a multi-class SVM model based on the DAG principle, and proposes an improved adaptive particle swarm optimization (APSO) to train the SVM model. Then we apply the proposed method to the voltage telemetry data set of the satellite momentum wheel. Compared with other methods, the results show that the proposed method has a good performance in improving the recognition accuracy and detection rate, and it can also effectively reduce the false alarm rate and the missed alarm rate. Therefore, the method proposed in this paper has a good development prospect in the field of anomaly detection of spacecraft.

In the future work, more real-world datasets will be applied to verify the effectiveness of the detection ability of the proposed method. In addition, more methods based on artificial neural networks will be studied to further improve the versatility of anomaly detection methods.

Author Contributions: Conceptualization, methodology and validation, Y.L. and M.L.; investigation, P.L. and R.W.; data curation and project administration, M.X.; writing—original draft preparation, M.L.; writing—review and editing, Y.L. All authors have read and agreed to the published version of the manuscript.

Funding: This research was funded by the National Natural Science Foundation of China (No.52075117), the Science Research Project (JSZL2020203B004) and the Key Laboratory Opening Funding of Harbin Institute of Technology (HIT.KLOF.2016.077, HIT.KLOF.2017.076, HIT.KLOF. 2018.074 and HIT.KLOF. 2018.076).

Institutional Review Board Statement: Not applicable.

Informed Consent Statement: Not applicable.

Data Availability Statement: The data included in this study are all owned by the research group and will not be transmitted.

Conflicts of Interest: The authors declare no conflict of interest.

References

1. Zhuang, M.; Tan, L.; Song, S. Fixed-time attitude coordination control for spacecraft with external disturbance. *ISA Trans.* **2021**, *114*, 150–170. [CrossRef] [PubMed]
2. Luo, Y.; Cui, L.; Zhang, J.; Ma, J. Vibration mechanism and improved phenomenological model of the planetary gearbox with broken ring gear fault. *J. Mech. Sci. Technol.* **2021**, *35*, 1867–1879. [CrossRef]
3. Li, Z.; Cheng, Y.; Wang, H.; Wang, H. Fault detection approach applied to inertial navigation system/air data system integrated navigation system with time-offset. *IET Radar Sonar Navig.* **2021**, *15*, 945–956. [CrossRef]

4. Zhang, W.; Zhou, J. Fault Diagnosis for Rolling Element Bearings Based on Feature Space Reconstruction and Multiscale Permutation Entropy. *Entropy* **2019**, *21*, 519. [CrossRef] [PubMed]
5. Wang, X.; Si, S.; Li, Y. Multiscale Diversity Entropy: A Novel Dynamical Measure for Fault Diagnosis of Rotating Machinery. *IEEE Trans. Ind. Inform.* **2021**, *17*, 5419–5429. [CrossRef]
6. Wodecki, J. Time-Varying Spectral Kurtosis: Generalization of Spectral Kurtosis for Local Damage Detection in Rotating Machines under Time-Varying Operating Conditions. *Sensors* **2021**, *21*, 3590. [CrossRef]
7. Cai, G.; Yang, C.; Pan, Y. EMD and GNN-AdaBoost fault diagnosis for urban rail train rolling bearings. *Discret. Contin. Dyn. Syst. S* **2019**, *12*, 1471–1487. [CrossRef]
8. Neupane, D.; Seok, J. Bearing Fault Detection and Diagnosis Using Case Western Reserve University Dataset with Deep Learning Approaches: A Review. *IEEE Access* **2020**, *8*, 93155–93178. [CrossRef]
9. Hu, Q.; Zhang, X.; Niu, G. Observer-based fault tolerant control and experimental verification for rigid spacecraft. *Aerosp. Sci. Technol.* **2019**, *92*, 373–386. [CrossRef]
10. Hou, S.; Sun, H.; Li, Q.; Tang, X. Design and experimental validation of a disturbing force application unit for simulating spacecraft separation. *Aerosp. Sci. Technol.* **2021**, *113*, 106674. [CrossRef]
11. Song, B.-P.; Zhou, R.-D.; Yang, X.; Zhang, S.; Yang, N.; Fang, J.-Y.; Song, F.-L.; Zhang, G.-J. Surface electrostatic discharge of charged typical space materials induced by strong electromagnetic interference. *J. Phys. D Appl. Phys.* **2021**, *54*, 275002. [CrossRef]
12. Boone, N.R.; Bettinger, R.A. Spacecraft survivability in the natural debris environment near the stable Earth-Moon Lagrange points. *Adv. Space Res.* **2021**, *67*, 2319–2332. [CrossRef]
13. McGarry, J.F.; Carabajal, C.C.; Saba, J.L.; Reese, A.R.; Holland, S.T.; Palm, S.P.; Swinski, J.A.; Golder, J.E.; Liiva, P.M. ICESat-2/ATLAS Onboard Flight Science Receiver Algorithms: Purpose, Process, and Performance. *Earth Space Sci.* **2021**, *8*, 4. [CrossRef]
14. Kumar, R.R.; Cirrincione, G.; Cirrincione, M.; Tortella, A.; Andriollo, M. Induction Machine Fault Detection and Classification Using Non-Parametric, Statistical-Frequency Features and Shallow Neural Networks. *IEEE Trans. Energy Convers.* **2020**, *36*, 1070–1080. [CrossRef]
15. Tao, L.; Yang, X.; Zhou, Y.; Yang, L. A Novel Transformers Fault Diagnosis Method Based on Probabilistic Neural Network and Bio-Inspired Optimizer. *Sensors* **2021**, *21*, 3623. [CrossRef] [PubMed]
16. Lin, Y.; Ge, H.; Chen, S.; Pecht, M. Two-level fault diagnosis RBF networks for auto-transformer rectifier units using multi-source features. *J. Power Electron.* **2020**, *20*, 754–763. [CrossRef]
17. Wang, T.; Wang, J.; Wu, Y.; Sheng, X. A fault diagnosis model based on weighted extension neural network for turbo-generator sets on small samples with noise. *Chin. J. Aeronaut.* **2020**, *33*, 2757–2769. [CrossRef]
18. Dong, H.; Chen, F.; Wang, Z.; Jia, L.; Qin, Y.; Man, J. An Adaptive Multisensor Fault Diagnosis Method for High-Speed Train Traction Converters. *IEEE Trans. Power Electron.* **2021**, *36*, 6288–6302. [CrossRef]
19. Belagoune, S.; Bali, N.; Bakdi, A.; Baadji, B.; Atif, K. Deep learning through LSTM classification and regression for transmission line fault detection, diagnosis and location in large-scale multi-machine power systems. *Measurement* **2021**, *177*, 109330. [CrossRef]
20. Oh, S.; Han, S.; Jeong, J. Multi-Scale Convolutional Recurrent Neural Network for Bearing Fault Detection in Noisy Manufacturing Environments. *Appl. Sci.* **2021**, *11*, 3963. [CrossRef]
21. Lv, X.; Wang, H.; Zhang, X.; Liu, Y.; Jiang, D.; Wei, B. An evolutional SVM method based on incremental algorithm and simulated indicator diagrams for fault diagnosis in sucker rod pumping systems. *J. Pet. Sci. Eng.* **2021**, *203*, 108806. [CrossRef]
22. Shi, Q.; Zhang, H. Fault Diagnosis of an Autonomous Vehicle with an Improved SVM Algorithm Subject to Unbalanced Datasets. *IEEE Trans. Ind. Electron.* **2021**, *68*, 6248–6256. [CrossRef]
23. Han, T.; Zhang, L.; Yin, Z.; Tan, A.C. Rolling bearing fault diagnosis with combined convolutional neural networks and support vector machine. *Measurement* **2021**, *177*, 109022. [CrossRef]
24. Lu, Y.; Li, Y. A novel data-driven method for maintenance prioritization of circuit breakers based on the ranking SVM. *Int. J. Electr. Power Energy Syst.* **2021**, *129*, 106779. [CrossRef]
25. Liu, Q.; Liu, W.; Mei, J.; Si, G.; Xia, T.; Quan, J. A New Support Vector Regression Model for Equipment Health Diagnosis with Small Sample Data Missing and Its Application. *Shock. Vib.* **2021**, *2021*, 6675078. [CrossRef]
26. Cuong-Le, T.; Nghia-Nguyen, T.; Khatir, S.; Trong-Nguyen, P.; Mirjalili, S.; Nguyen, K.D. An efficient approach for damage identification based on improved machine learning using PSO-SVM. *Eng. Comput.* **2021**, *20*, 1–16. [CrossRef]
27. Wang, B.; Zhang, X.; Xing, S.; Suna, C.; Chena, X. Sparse representation theory for support vector machine kernel function selection and its application in high-speed bearing fault diagnosis. *ISA Trans.* **2021**, *4*, 60. [CrossRef]
28. Zheng, H.; Wang, R.; Yang, Y.; Li, Y.; Xu, M. Intelligent fault identification based on multisource domain generalization towards actual diagnosis scenario. *IEEE Trans. Ind. Electron.* **2020**, *67*, 1293–1304. [CrossRef]
29. Yan, X.; Liu, Y.; Ding, P.; Jia, M. Fault Diagnosis of Rolling-Element Bearing Using Multiscale Pattern Gradient Spectrum Entropy Coupled with Laplacian Score. *Complexity* **2020**, *2020*, 4032628. [CrossRef]
30. Song, Y.; Si, W.; Dai, F.; Yang, G. Weighted ReliefF with threshold constraints of feature selection for imbalanced data classification. *Concurr. Comput. Pract. Exp.* **2020**, *32*, 14. [CrossRef]
31. Richman, J.; Lake, D.; Moorman, J. Sample entropy. *Methods Enzymol.* **2004**, *384*, 172–184. [PubMed]
32. Costa, M.; Goldberger, A.L.; Peng, C.-K. Multiscale Entropy Analysis of Complex Physiologic Time Series. *Phys. Rev. Lett.* **2002**, *89*, 068102. [CrossRef] [PubMed]

33. Huffman, D. A Method for the construction of minimum-redundancy codes. *Proc. IRE* **1952**, *40*, 1098–1101. [CrossRef]
34. Kennedy, J.; Eberhart, R. Particle swarm optimization. In Proceedings of the ICNN'95—International Conference on Neural Networks, Perth, WA, Australia, 27 November–1 December 1995; Volume 4, pp. 1942–1948. [CrossRef]

Article

A New Universal Domain Adaptive Method for Diagnosing Unknown Bearing Faults

Zhenhao Yan [1], Guifang Liu [1,*], Jinrui Wang [1], Huaiqian Bao [1], Zongzhen Zhang [1,2], Xiao Zhang [1] and Baokun Han [1]

1. College of Mechanical and Electronic Engineering, Shandong University of Science and Technology, Qingdao 266590, China; 201982050024@sdust.edu.cn (Z.Y.); wangjinrui@sdust.edu.cn (J.W.); bhqian@sdust.edu.cn (H.B.); skd996576@sdust.edu.cn (Z.Z.); 201982050027@sdust.edu.cn (X.Z.); hanbaokun@sdust.edu.cn (B.H.)
2. College of Energy and Power Engineering, Nanjing University of Aeronautics and Astronautics, Nanjing 210016, China
* Correspondence: skd994188@sdust.edu.cn; Tel.: +86-137-3098-1732

Abstract: The domain adaptation problem in transfer learning has received extensive attention in recent years. The existing transfer model for solving domain alignment always assumes that the label space is completely shared between domains. However, this assumption is untrue in the actual industry and limits the application scope of the transfer model. Therefore, a universal domain method is proposed, which not only effectively reduces the problem of network failure caused by unknown fault types in the target domain but also breaks the premise of sharing the label space. The proposed framework takes into account the discrepancy of the fault features shown by different fault types and forms the feature center for fault diagnosis by extracting the features of samples of each fault type. Three optimization functions are added to solve the negative transfer problem when the model solves samples of unknown fault types. This study verifies the performance advantages of the framework for variable speed through experiments of multiple datasets. It can be seen from the experimental results that the proposed method has better fault diagnosis performance than related transfer methods for solving unknown mechanical faults.

Keywords: fault diagnosis; rotating machinery; transfer learning; domain adaptation

1. Introduction

Existing deep neural networks have shown superior performance in various diagnostic tasks for rotating component faults due to their impressive feature learning capabilities [1–3]. Such networks include the convolutional neural network [4,5], recurrent neural network [6], and restricted Boltzmann machine [7]. The outstanding performances of these networks heavily depend on the pretraining of deep diagnostic networks with real sample data from the same domain as the test data [8]. However, under actual operating conditions, the dataset is often time-varying and unknowable. Improving the generalization capability of a model under variable working conditions has been regarded as a potential solution for solve unknown working conditions.

Domain discrepancy causes the model based on the previous training data to perform poorly with the new test data set [9,10]. The typical solution to this problem is to pre-train the model and fine-tune the diagnostic network trained from the source domain with the feature distribution of the target domain [11], and the method for the marginal distribution alignment of feature spaces is widely used to narrow the distance between two different domains [12]. Li et al. [13] proposed a fault diagnosis model based on multi-scale permutation entropy (MPE) and multi-channel fusion convolutional neural networks (MCFCNN), which constructs a feature vector set by permuting entropy so that the high accuracy and stability of fault diagnosis are realized. Guo et al. [14] reported a new transfer learning network, which gradually realized the multi-module operation of automatic features learning and machine health status recognition through a one-dimensional convolutional network. Singh et al. [15] presented a deep convolution model to diagnose the type of the gearbox fault under the obvious change of speed. The model minimizes the cross-entropy loss of the source domain and the maximum mean discrepancy loss between the two domains to obtain superior diagnostic performance. Hasan et al. [16] proposed a transfer diagnosis framework based on high-order spectral analysis and multitask learning, which can diagnose non-stationary and non-linear rolling bearing signals in combination with different modes of a given fault type. As can be seen from the above-mentioned networks, solving the problem of domain discrepancies has become a tacit prerequisite for current fault diagnosis.

Traditional diagnostic networks usually assume that the label space of the fault samples in the target domain and the source domain is consistent. However, in actual engineering practice, the fault type of the target domain is often difficult to predict, and the fault type label space is often smaller than the source domain fault label space. Therefore, Cao et al. [17] proposed the use of selective weighting to maximize the positive migration of shared tag space data; this approach can achieve the purpose of per-class adversarial distribution matching. Zhang et al. [18] established an importance weighted adversarial network. This network is especially suitable for partial domain adaptation where the number of fault types in the target domain is less than the number of fault types in the source domain, and can effectively reduce the distribution difference to realize knowledge migration and the fault diagnosis of the target sample. Li et al. [19] suggested applying unsupervised prediction consistency schemes and conditional data alignment for partial domain adaptation. This method effectively solves the partial domain adaptation problem that the target domain data under unsupervised training cannot cover the entire healthy label space. Jia et al. [20] proposed a weighted subdomain adaptation network (WSAN), and a weighted local maximum-mean-discrepancy (WLMMD) is introduced to obtain the transferable information and weight of the sample to realize the diagnosis of the fault type. The research on partial domain adaptation pushes the field of intelligent fault diagnosis into a practical setting.

However, only a very small number of networks can cope with the identification and diagnosis of sudden unknown fault types in the existing fault diagnosis models. We cannot know that the fault type of the target samples must belong to the source domain label space when providing unlabeled target samples. Therefore, open set recognition is an urgent problem faced by transfer learning to broaden practical application scenarios. Busto et al. [21] were the first to suggest marking the shared classes of the source and target domains as general classes and constructed an iterative method to solve the labeling problem. Saito et al. [22] modified the description of open set domain adaptation, which allows only the target domain to contain the private label set. His team also added a boundary between the source domain and the target domain to facilitate the separation of unknown fault samples from known fault samples. This method has been widely evaluated in the field. You et al. [23] provided a concept of universal domain in the field of image recognition which allows intersection between source and target domains and provides a benchmark for future related research.

Considering that the current domain transfer methods often assume that the fault type of the test data is the same as the training data set, while ignores that the specific working conditions and label types of the target domain samples are often unpredictable. It is impossible to diagnose the fault type by directly comparing the distribution of the source domain and the target domain. Thus, we propose a new universal domain adaptation (UDA) method for fault diagnosis under the changing conditions of bearing speed. As shown in Figure 1, the model allows different types of faults to exist between data sets and generates a feature center belonging to each fault type for fault diagnosis by learning the fault features of each fault sample. In order to solve the problem of negative model transfer caused by the input of unknown samples into the network, the model proposes three optimization goals, and train the network gradually by optimizing the objective function to alleviate the phenomenon of negative network transfer. The main contributions of this model are as follows:

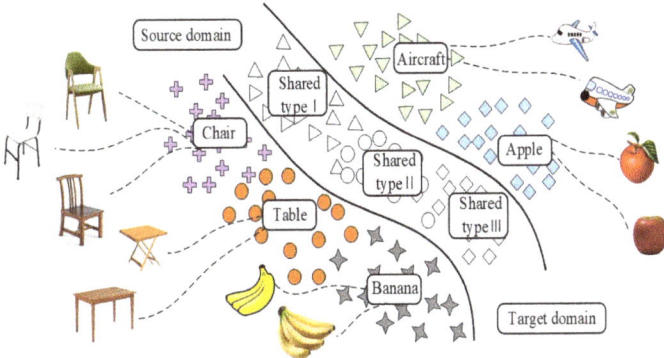

Figure 1. Universal domain adaptation setting (unshaded shapes indicate shared labels).

1. The proposed model breaks the assumption of the shared label space in the field of mechanical fault diagnosis and proposes the universal domain to solve the fault type samples that did not appear in the training dataset.
2. The proposed network innovatively proposes to rely on source domain samples to generate feature centers of each fault type and determine the fault type based on the distance between the feature extracted from the sample and the feature center.
3. The model introduces Wasserstein distance to measure the marginal probability distribution between different data, and three optimization equations are added to the network training to optimize the model to alleviate the negative transfer problem of the network when solving unknown domains.

In this paper, a new transfer learning model based on universal domain adaptation is proposed and the proposed model is described in detail. The specific article structure is organized as follows. The details of the proposed method for fault diagnosis under changing speed conditions are provided in Section 2. The fault diagnosis experiment with two sets of bearing data is presented in Section 3. Finally, the conclusions are provided in Section 4.

2. Research Methods
2.1. Proposed Framework

The frame structure of the proposed approach is shown in Figure 2.

The proposed framework adopts two modules, i.e., the feature extractor G and the classifier C. The feature extractor G is composed of 4 fully connected layers, and the dimensions of the samples extracted from each layer are 512, 128, 64, and 16 dimensions. The classifier C is a two-layer Softmax classifier, which is used to diagnose fault sample types. The original source time-domain signal is processed by FFT and input into the feature extractor G to extract the features of the source domain fault sample. The extracted fault features are then classified by the classifier C to extract the feature center of each fault type from the feature signal of the source domain gradually. After the first model pre-classification, the features of the target domain fault samples are added multiple times with tiny noise containing their own features, and the distance from the feature center is measured. The model realizes the fault diagnosis of the target domain samples after many times of learning and training. The training process of the model is described in detail in Section 3.2. The model introduces the following target objects to improve the diagnostic performance and generalization ability.

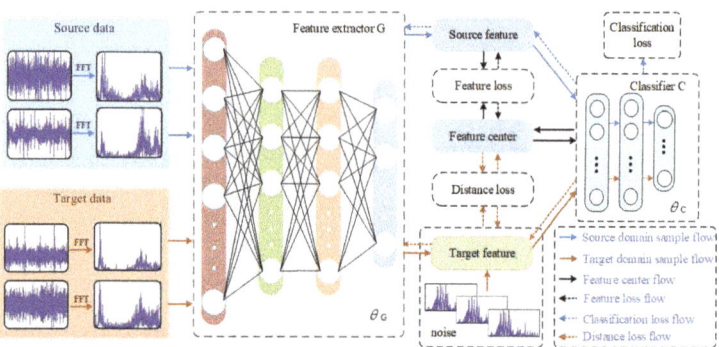

Figure 2. Framework diagram of the proposed method.

2.1.1. Classification Loss

Minimizing the classification error of source domain samples is the first optimization goal of the proposed framework. The classifier learns classification knowledge from the labeled samples in the source domain. The standard Softmax regression loss is selected as the objective function [24]. The specific function formula and explanation are as follows:

$$L_C = -\frac{1}{m}[\sum_{i=1}^{m}\left(1-y^{(i)}\right)\log\left(1-h_\theta\left(x^{(i)}\right)\right)+\sum_{i=1}^{m}y^{(i)}\log h_\theta\left(x^{(i)}\right)] \quad (1)$$

where $x^{(i)}$ and $y^{(i)}$ represent the input signal of the i-th sample and the probability output corresponding to the sample, and $h_\theta(x^{(i)})$ is the set of probabilities of various fault types corresponding to the i-th sample.

2.1.2. Feature Loss

Feature loss is used to correct the error loss caused by discarding useless fault type features in the process of extracting feature centers. Feature loss can be expressed as the absolute difference between the feature extracted from the source domain and the feature center generated by the learning process. The function formula [25] is as follows:

$$L_F(x_c, o_c) = \frac{1}{C}\sum_{c=1}^{C}|f(x_c) - f(o_c)| \quad (2)$$

where x_c and o_c are the c-th features of the feature extraction and feature center.

2.1.3. Distance Loss

Wasserstein distance, which is often used to measure the discrepancy between different distributions, can be understood as the minimum consumption under optimal path planning. The Wasserstein distance is used as distance loss to reflect accurately the distance between the two distributions with little to no overlap in the support set with the objective of measuring the overall distance between the feature center and the target domain feature [26]. The function formula is as follows:

$$L_D(P_1, P_2) = \inf_{\gamma \in \prod(p_1, p_2)} E_{(x,y) \sim \gamma}[\|x - y\|] \quad (3)$$

where $\prod(p_1, p_2)$ is the set of all possible joint distributions that combine the P_1 and P_2 distributions, and γ represents the joint distribution of each possible fault type. x represents the feature center sample feature and y represents the target sample feature.

2.2. Training Process

The goal of the model is to identify the fault type of the target domain sample and reduce the domain distance between two identical faults. At the same time, the fault types of unknown samples in the target domain are identified. The training process of the model is shown in the figure below:

Step 1: Figure 3 shows that the network learns the features of the fault types from the source domain samples to form the characteristic centers of multiple fault types. The classifier tries to pre-classify the target domain samples and attempts to shorten the distance discrepancy between domains. Therefore, the source domain classification loss is introduced into the model. The mathematical equation used is as follows:

$$L_C = -\frac{1}{m}\left[\sum_{i=1}^{m}\sum_{j=1}^{k}I\{y^{(i)}=j\}\log\frac{\exp(\theta_j^T \cdot x^{(i)})}{\sum_{l=1}^{k}\exp(\theta_l^T \cdot x^{(i)})}\right] \quad (4)$$

where m represents the number of samples in the source domain and $I[\cdot]$ is an index function used to represent the value of the probability that the sample is true. $\theta_1, \theta_2, ..., \theta_k \in \Re^{n+1}$ are the parameters of the model and $1/\left(\sum_{j=1}^{k}e^{\theta_j^T x^{(i)}}\right)$ normalizes the distribution such that it sums to 1.

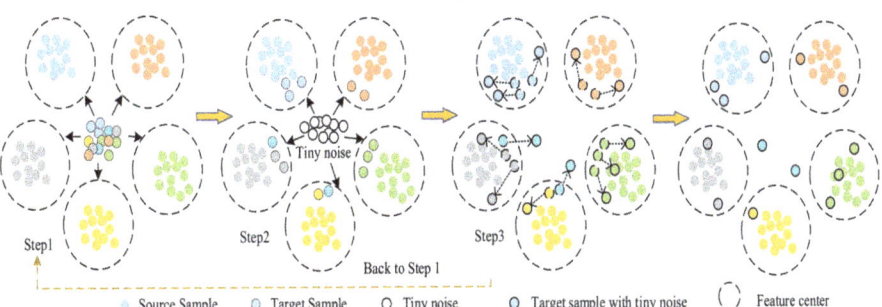

Figure 3. Model training steps.

Step 2: Tiny noise is mixed into the target domain samples in the classification, and these tiny noises merge the features of the target domain sample extracted from the feature extractor G. The fault features of the target domain samples mixed with the noise will then undergo a slight change. Given that the mixed tiny noise is related to the target sample features themselves, the extracted fault sample feature will be closer to the feature center of its own fault type. The function formula is as follows:

$$X_{to} = X_t + \lambda \cdot \widehat{x} \cdot t \quad (5)$$

where λ represents the feature coefficient of the tiny noise. $\widehat{x}$ is the feature coefficient of the extracted target domain, which is the sample feature extracted from the target domain sample. o^t is Gaussian white noise used as the tiny noise for network training.

Step 3: The network recalculates the distance between the features of the target sample and the center of each fault type after the addition of noise, and the distance loss between the features of the target sample and the feature center is calculated to judge the fault. The specific distance loss function [25] is shown in the following formula:

$$L_{dis} = \left|\frac{1}{m^s}\sum_{i=1}^{m^s}T(x_{si}) - \frac{1}{m^t}\sum_{i=1}^{m^t}T(x_{ti})\right| \quad (6)$$

where x_{si} and x_{ti} are the i-th features extracted from the target domain X_t and source domain X_s through the fully connected layer.

The three steps of model training are looped continuously until the expected performance is achieved as shown in Figure 3. The network repeatedly adds tiny noise interference containing the characteristics of the target domain sample to the target domain samples and measures the feature distance to ensure the accurate diagnosis of the fault type of the target domain samples. The stable samples that have been classified accurately do not undergo classification changes after multiple small disturbances are added, whereas the active samples that have been classified incorrectly jump or leave the feature center.

3. Experimental Verification

3.1. Experimental Dataset Description

The intelligent fault diagnosis methods trained with the labeled data are required to classify the unlabeled data accurately to validate the effectiveness of this method in universal domain transfer learning. Therefore, as discussed in this section, the datasets acquired from two dedicated rotating part workbenches are used for bearing fault diagnosis experiments.

CWRU: The Case Western Reserve University (CWRU) bearing dataset was collected from an experimental platform provided by the CWRU [27]. The CWRU workbench collected sample data of four health conditions at the 6 o'clock position (orthogonal area of applied load) of the deep groove ball bearing on the drive end of the motor housing. The four health conditions were normal condition (NC), inner race failure (IF), outer race failure (OF), and ball failure (BF). The sampling frequency at the time of data collection is set to 48 kHz, and each fault type was run with varying degrees of damage (0.007-, 0.014-, and 0.021-inch fault diameters). Each type of fault data was collected by the test motor running at three different motor speeds (i.e., 1772, 1750, and 1730 rpm) for fault diagnosis. The CWRU dataset information is shown in Table 1.

Table 1. Information of the two datasets.

Dataset	Class Label	1	2	3	4	5	6	7	8	9	10
CWRU	Fault location	N/A	IF	IF	IF	BF	BF	BF	OF	OF	OF
	Fault size (mil)	0	7	14	21	7	14	21	7	14	21
SDUST	Fault location	IF	IF	N/A	OF	OF	RF	RF	ROF	ROF	
	Fault size (mm)	0.2	0.4	0	0.2	0.4	0.2	0.4	0.2	0.4	

SDUST: The Shandong University of Science and Technology (SDUST) bearing dataset was collected from a diagnostic test bench specially designed for bearing faults. The time-domain signal of bearings at different speeds of the motor is collected. Figure 4a shows that the bearing fault test bench is composed of a motor, a rotor, a brake, a bearing seat, and two shaft couplings. The cylindrical roller bearing faults in the SDUST dataset contains three single types of fault and a type of composite fault, which are: OF, roller fault (RF), IF, and roller and outer race fault (ROF). Figure 4 is a schematic diagram of the three single types of fault bearings. The collected bearing signals of each type of fault are divided into two fault severity levels: crack 0.2 mm and crack 0.4 mm. The NC time-domain signal was added to the SDUST dataset to obtain nine types of faults as shown in Table 1. Each acceleration sensor was respectively installed in different parts of the bearing seat, and the sampling frequency was set at 25.6 kHz. The motor speed was set to 1500 r/min, 1800 r/min, and 2000 r/min during data collection, and a total of 200 samples were collected for each fault health status at different motor speeds, and each time-domain sample contained 6400 data points.

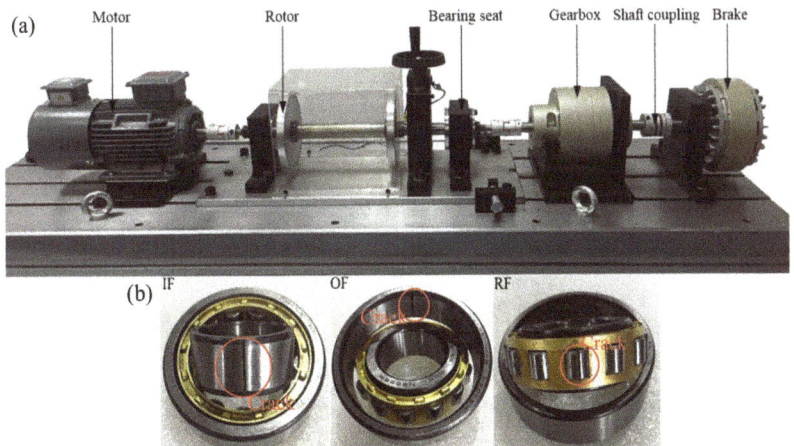

Figure 4. (a) Bearing fault test rig and (b) three single types of fault bearings.

3.2. Compared Methods Description

The proposed method shares the same experimental configuration and test dataset with all the following comparison methods to evaluate the diagnostic advantages of the proposed model.:

Baseline: First, a baseline method without a special technique is proposed to diagnose the UDA problem [28]. The feature extractor and classifier are trained under supervision, and the network is used directly for the fault diagnosis of the target domain dataset.

L1/2-SF: The L1/2-SF (L1/2 regularized sparse filtering) approach [29] is widely used as an excellent method for bearing and gear fault diagnosis. This method follows the traditional unsupervised machine learning model, thus providing a benchmark for the proposed method.

WD-MCD: The Wasserstein Distance—Maximum Classifier Discrepancy (WD–MCD) approach maximizes the output discrepancy of the classifier and combines marginal probability distribution adaptation to focus on the diagnosis of the transfer model [20]. This model is compared with the proposed model because it is a popular transfer learning method.

BN–SAE: As a popular method, the Batch Normalization—Stacked AutoEncoders (BNSAE) approach [30] is an adaptive reparametrization algorithm that aims not to optimize but to regularize the model. The effect of the healthy data classification scheme can thus be examined.

3.3. Experimental Results Display

As reported in this section, the UDA problem is experimentally verified. The model parameters during the experiment are shown in Table 2 and are mainly determined in accordance with the verification results of the diagnostic task.

Table 2. Parameters used in this study.

Parameter	Value	Parameter	Value
Epochs in general training E_s	2000	Sample dimension S	1200
Epochs in testing E_t	500	Feature center dimension C_d	16
Batch size B_s	10	First-level feature dimension f_1	512
Dropout_rate D	0.1	Second-level feature dimension f_2	128
Learning rate Lr	0.001	the feature coefficient of the tiny noise λ	0.05

The detailed task information is shown in Table 3. This information randomly selects the fault type for fault diagnosis in the two datasets. In order to verify the fault diagnosis performance of the framework under variable speed conditions, the experiment sets the fault type diagnosis in the phase of gradually increasing bearing speed and the fault type diagnosis in the phase of gradually decreasing speed. The "source classes" represent the fault type of the source domain training sample, the type of fault marked red represents the private type of the source domain (the category does not appear in the target domain fault sample), and the "unknown class" represents the fault type that has not been learned during source domain training. A total of 100 labeled samples under each machine condition are randomly selected as source domain data for model training, and 100 unlabeled samples are used as target domain samples for experimental verification. An average of 10 experiments for each group of results is performed to reduce the effect of randomness.

Table 3. Universal domain transfer learning task information.

CWRU				SDUST			
Task Name	Transfer (Speed)	Source Classes	Unknown Class	Task Name	Transfer (Speed)	Source Classes	Unknown Class
A1	1730 → 1750	1, 2, 3, 5, 8	6	B1	1500 → 1800	1, 3, 4, 6, 7	8
A2	1730 → 1750	1, 2, 3, 6, 9	8	B2	1500 → 1800	2, 3, 4, 5, 7	9
A3	1730 → 1750	1, 4, 5, 7, 10	2	B3	1500 → 1800	1, 3, 5, 6, 8	7
A4	1730 → 1750	No unknown fault		B4	1500 → 1800	No unknown fault	
A5	1730 → 1772	1, 2, 3, 5, 8	6	B5	1500 → 2000	1, 3, 4, 6, 7	8
A6	1730 → 1772	2, 3, 5, 7, 9	1	B6	1500 → 2000	1, 3, 4, 5, 6	8
A7	1730 → 1772	4, 5, 6, 8, 9	3	B7	1500 → 2000	1, 3, 5, 6, 8	2
A8	1750 → 1730	1, 2, 3, 5, 8	6	B8	1800 → 1500	1, 3, 4, 6, 7	8
A9	1750 → 1730	1, 4, 5, 7, 10	2	B9	1800 → 1500	1, 2, 3, 4, 8	6
A10	1772 → 1730	1, 2, 4, 5, 8	10	B10	2000 → 1500	1, 3, 4, 6, 8	9

3.3.1. CWRU Task Set Result Analysis

Table 4 shows the fault diagnosis results of the proposed model using the CWRU dataset for different universal domain tasks. The selected comparison approaches are currently the highly popular domain adaptive and transfer learning methods. The stable diagnostic performance presented by the proposed method in the CWRU dataset task group shows its superiority in solving universal domain adaptation. Furthermore, the proposed method generally obtains smaller standard deviations than other models when performing different tasks, indicating that it has good convergence in the experimental process.

Table 4. Means of the testing accuracies in different tasks with the CWRU dataset (%).

Method	Baseline	L1/2-SF	WD-MCD	DA-BNSAE	Proposed
A1	54.67 (±12.4)	82.2 (±6.3)	80.08 (±2.6)	78.23 (±0.4)	82.98 (±2.1)
A2	43.2 (±8.4)	73.68 (±5.1)	75.02 (±3.2)	74.22 (±2.1)	80.32 (±1.4)
A3	57.46 (±3.7)	70.31 (±8.2)	82.38 (±5.0)	73.65 (±3.4)	90.31 (±2.7)
A4	72.56 (±6.2)	98.83 (±1.7)	99.43 (±0.5)	98.53 (±1.6)	99.88 (±0.2)
A5	55.07 (±4.6)	69.45 (±1.2)	80.2 (±0.4)	71.07 (±2.7)	80.96 (±6.5)
A6	47.9 (±3.0)	69.67 (±3.2)	72.15 (±3.8)	65.2 (±2.3)	77.38 (±9.3)
A7	33.3 (±6.4)	64.78 (±11.7)	73.06 (±5.4)	64.93 (±15.1)	92.37 (±2.2)
A8	48.47 (±10.7)	74.32 (±3.6)	79.6 (±3.3)	75.54 (±6.5)	84.31 (±3.8)
A9	48.78 (±7.1)	71.56 (±5.9)	70.34 (±8.4)	67.3 (±7.9)	88.01 (±4.5)
A10	44.7 (±15.3)	55.62 (±5.5)	71.62 (±7.8)	65.07 (±13.6)	85.96 (±7.4)
Average	50.61	73.04	78.39	73.37	86.25

The L1/2-SF, the WD-MCD, and the DA-BNSAE methods achieve relatively ideal accuracy rates in the tasks with the CWRU dataset compared with the baseline method, and their accuracy in some tasks is as high as 80 or more. These methods generally have good feature recognition capabilities in the diagnosis of minor faults. However, in serious fault diagnosis, the proposed model has superior feature recognition and diagnosis performance.

Typical task cases are represented by tasks A3 and A7. The accuracy of the proposed method is as high as over 90%, and high accuracy is also obtained in the process of solving task 9. Task 4 is specially set as the task benchmark, which has no unknown fault type and belongs to pure rotation speed transfer. It can be seen that the proposed method can still guarantee high diagnostic accuracy. Comparing the standard deviation of each method reveals that the proposed method exhibits relatively stable performance in multiple tasks, further verifying its convergence performance. Although the performance in minor fault diagnosis shown by the proposed method is not as good as that of the three comparison methods, the superiority of the modified model can be seen in the overall task performance comparison. Moreover, the proposed model is more suitable for fault diagnosis problems in the universal domain than other comparison methods.

The t-distributed Stochastic Neighbor Embedding (t-SNE) method [31] is widely used in the display of various fault diagnosis results. This approach can reduce the dimensionality of the output of the high-dimensional features by the model and provide visualization processing. The A1 task based on the CWRU dataset is selected as the demonstration experiment for t-SNE dimensionality reduction processing for visually displaying the fault diagnosis performance of each network as shown in Figure 5.

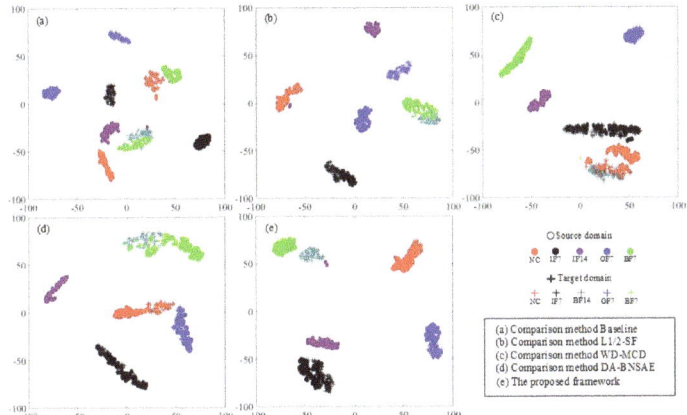

Figure 5. Feature visualization of the t-SNE results for the CWRU dataset in the A1 task.

The baseline approach exhibits poor fault diagnosis performance as illustrated in Figure 5a. It not only fails to aggregate various types of faults completely, but it also clusters BF14 faults and BF7 faults together. As can be seen in Figure 5b,d, the L1/2-SF, and the DA-BNSAE approach incorrectly classify the BF14 fault as the BF7 fault, and the L1/2-SF approach also shows that the target domain and the source domain samples of the OF7 fault are not clustered. As presented in Figure 5c, although the WD-MCD approach has a better fault clustering effect than the previous three methods, a situation wherein the fault type BF14 is mistakenly classified as a healthy sample exists. The clustering dimensionality reduction graph of the proposed method is shown in Figure 5e. Although a small number of IF14 fault samples are close to the BF14 fault in the proposed method, the target domain and the source domain samples of various types of faults have obvious domain boundaries and show a good fault classification effect.

Figure 6 depicts that the training accuracy and testing accuracy of the method tends to stabilize as the training progresses, and feature and distance losses in the model gradually decrease as accuracy increases.

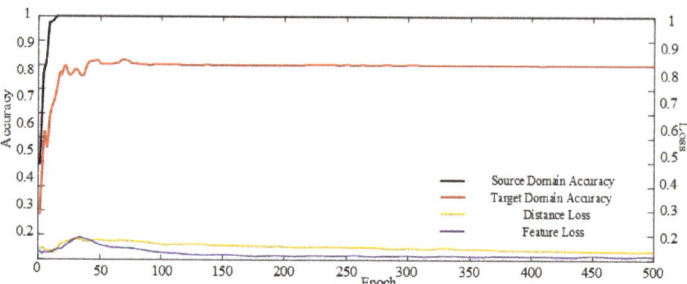

Figure 6. Accuracy and training loss of the proposed method in the A1 task.

3.3.2. SDUST Task Set Result Analysis

The experimental accuracy for the SDUST dataset is shown in Table 5. Given that this dataset has more drastic speed changes, diagnostic performance with this dataset is worse than that with the CWRU dataset. It can be seen that the three comparison methods show good diagnostic performance in individual tasks. However, the proposed method is more convincing in terms of overall fault classification effect and stability. The model still provides superior fault diagnosis accuracy under variable speed conditions that further validates its robustness and superiority for UDA problems.

Table 5. Means of the testing accuracies in different tasks with the SDUST dataset (%).

Task	Baseline	L1/2-SF	WD-MCD	DA-BNSAE	Proposed
B1	44.8 (±3.4)	79.44 (±4.6)	77.32 (±2.7)	77.18 (±3.9)	80.04 (±2.1)
B2	41.45 (±7.5)	72.72 (±2.4)	67.26 (±5.6)	68.67 (±7.2)	83.06 (±5.7)
B3	47.67 (±4.3)	71.32 (±8.5)	76.84 (±3.0)	76.57 (±2.9)	81.96 (±4.3)
B4	70.42 (±3.4)	97.53 (±1.4)	99.44 (±0.9)	98.94 (±1.1)	99.76 (±0.5)
B5	45.07 (±6.3)	75.97 (±3.2)	78.67 (±9.6)	64.87 (±4.1)	84.72 (±5.9)
B6	46.3 (±7.2)	67.05 (±10.5)	76.1 (±13.7)	58.31 (±17.3)	79.9 (±3.5)
B7	52.78 (±3.8)	72.14 (±7.2)	76.18 (±9.4)	72.23 (±4.5)	90.7 (±2.9)
B8	51.21 (±5.2)	78.3 (±5.9)	81.04 (±2.1)	64.01 (±7.9)	90.44 (±4.7)
B9	44.23 (±6.5)	49.52 (±17.8)	64.4 (±14.5)	60.03 (±15.9)	82.64 (±7.9)
B10	39.78 (±6.3)	57.23 (±5.6)	45.63 (±6.7)	57.22 (±9.0)	73.62 (±12.3)
Average	48.37	72.12	74.29	69.80	84.68

Considering that the experimental task is too heavy, the B1 task based on the SDUST dataset is selected as the demonstration experiment for t-SNE dimensionality reduction processing for visually displaying the fault diagnosis performance of each network (Figure 7). Although the baseline approach provides a good clustering of the source domain fault types, it shows a small amount of confusion between the fault type 3 (NC) and the fault type 8 (ROF0.2), as well as between the fault type 8 (ROF0.2) and the fault type 1 (IF0.2) in the target domain. The baseline approach mistakenly classifies the target domain fault type 8 (ROF0.2) sample as the fault RF0.2 as shown in Figure 7a. As can be seen from Figure 7c, the WD-MCD approach incorrectly classifies the NC samples at 1800 speed as the OF0.2 fault, and some samples as the fault type 8 (ROF0.2) in the target domain are mixed with the fault RF0.2. The DA-BNSAE network, one of the approaches used for comparison, confuses the fault type boundaries of IF0.2, RF0.2, and ROF0.2 faults as presented in Figure 7d. Comparing the proposed method with the L1/2-SF approach, it is found that although the L1/2-SF method has a good clustering effect on the source domain samples, there are still a small number of OF0.2, RF0.2, and ROF0.2 fault samples that are misclassified. The proposed network not only has a more obvious clustering effect on samples of various fault types but also has obvious separation between samples of different fault types, as shown in Figure 7b,e.

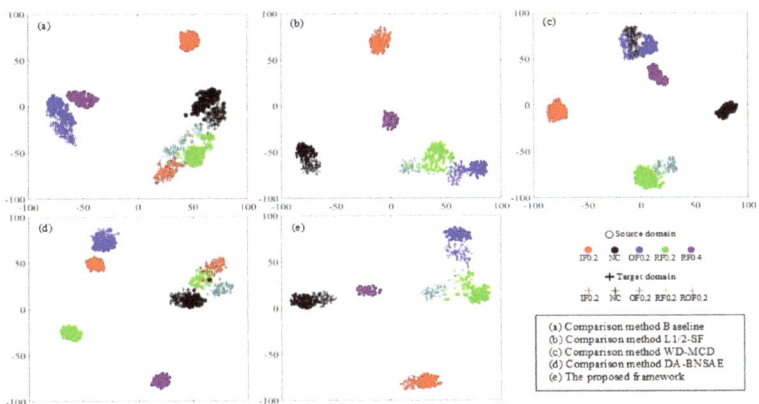

Figure 7. Feature visualization of the t-SNE results for the SDUST dataset in the B1 task.

Therefore, the proposed model can diagnose the samples of the unknown fault types more effectively than other networks.

4. Conclusions

This paper presents a new UDA method for bearing fault diagnosis under different working conditions, which breaks the assumption that the traditional domain adaptive network shares the label space and attempts to solve the unknown scale domain by using a universal label domain method. The proposed method was compared with the current popular domain adaptation methods in the experimental verification stage. Under the premise of sharing the experimental configuration and dataset, we set up multiple sets of experimental tasks for different actual work needs. Through multiple experimental verifications, it is concluded that the proposed method has higher classification accuracy and robustness than the comparison methods in diagnosing bearing datasets under variable conditions, and it can still guarantee high diagnostic performance even in the presence of bearing samples of unknown fault types. Therefore, the proposed method is more suitable for actual working conditions that change from time to time.

Author Contributions: Conceptualization, Z.Y. and J.W.; methodology, Z.Y.; software, H.B.; validation, Z.Z., X.Z., and B.H.; formal analysis, Z.Y.; investigation, Z.Y.; resources, J.W.; data curation, Z.Y.; writing-original draft preparation, Z.Y.; writing-review and editing, Z.Y.; visualization, Z.Y.; supervision, J.W.; project administration, G.L.; funding acquisition, J.W. All authors have read and agreed to the published version of the manuscript.

Funding: This research was supported by the National Natural Science Foundation of China, grant number 52005303, the Project of China Postdoctoral Science Foundation, grant number 2019M662399, and the Natural Science Foundation of Shandong Province, grant number ZR202020QE157.

Data Availability Statement: The data used to support the findings of this study are available from the corresponding author upon request.

Acknowledgments: This research was supported by special funds and experimental equipment from the Shandong University of Science and Technology.

Conflicts of Interest: The authors declare no conflict of interest.

Notations

B_s Batch size
C_d Feature center dimension
D Dropout_rate

E_s	Epochs in general training
E_t	Epochs in testing
f_1	First-level feature dimension
f_2	Second-level feature dimension
$h_\theta(x^{(i)})$	The probability set of various fault types for the i-th sample
$I[\cdot]$	An index function used to represent the value of the probability
Lr	Learning rate
m	The number of samples in the source domain
o^t	The tiny noise used for network training
o_c	The c-th features of the feature center
S	Sample dimension
$x^{(i)}$	The input signal of the i-th sample
$\hat{x}$	The feature coefficient of the extracted target domain
x_c	The c-th features of the feature extraction
x_{si}	The i-th features extracted from the target domain
x_{ti}	The i-th features extracted from the source domain
X_t	The target domain
X_s	The source domain
$y^{(i)}$	The probability output corresponding to the i-th sample
$\prod(p_1, p_2)$	The set of all possible joint distributions that combine the P_1 and P_2 distributions
$\theta_1, \theta_2, \dots, \theta_k$	The parameters of the model
θ_C	The parameters of the classifier
θ_G	The parameters of the feature extractor
λ	The feature coefficient of the tiny noise
γ	The joint distribution of each possible fault type

References

1. Jia, S.; Wang, J.; Zhang, X.; Han, B.K. A Weighted Subdomain Adaptation Network for Partial Transfer Fault Diagnosis of Rotating Machinery. *Entropy* **2021**, *23*, 424. [CrossRef]
2. Zhao, R.; Yan, R.; Chen, Z.; Mao, K.Z.; Wang, P.; Gao, R.X. Deep learning and its applications to machine health monitoring. *Mech. Syst. Signal Process* **2019**, *115*, 213–237. [CrossRef]
3. Liu, R.; Meng, G.; Yang, B.; Sun, C.; Chen, X. Dislocated time series convolutional neural architecture: An intelligent fault diagnosis approach for electric machine. *IEEE Trans. Ind. Inform.* **2017**, *13*, 1310–1320. [CrossRef]
4. Zhu, J.; Chen, N.; Peng, W. Estimation of bearing remaining useful life based on multiscale convolutional neural network. *IEEE Trans. Ind. Electron.* **2019**, *66*, 3208–3216. [CrossRef]
5. Lu, S.; Qian, G.; He, Q.; Liu, F. In Situ Motor Fault Diagnosis Using Enhanced Convolutional Neural Network in an Embedded System. *IEEE Sens. J.* **2020**, *20*, 8287–8296. [CrossRef]
6. Jiang, H.; Li, X.; Shao, H.; Ke, Z. Intelligent fault diagnosis of rolling bearings using an improved deep recurrent neural network. *Meas. Sci. Technol.* **2018**, *29*, 065107. [CrossRef]
7. Shao, H.; Jiang, H.; Li, X.; Liang, T.C. Rolling bearing fault detection using continuous deep belief network with locally linear embedding. *Comput. Ind.* **2018**, *96*, 27–39. [CrossRef]
8. Li, X.; Zhang, W.; Ding, Q. Cross-domain fault diagnosis of rolling element bearings using deep generative neural networks. *IEEE Trans. Ind. Electron.* **2018**, *66*, 5525–5534. [CrossRef]
9. Wang, H.; Li, S.; Song, L.; Cui, L. A novel convolutional neural network based fault recognition method via image fusion of multivibration-signals. *Comput. Ind.* **2019**, *105*, 182–190. [CrossRef]
10. Mao, W.; Sun, B.; Wang, L. A New Deep Dual Temporal Domain Adaptation Method for Online Detection of Bearings Early Fault. *Entropy* **2021**, *23*, 162. [CrossRef]
11. Jiao, J.; Zhao, M.; Lin, J.; Ding, C. Classifier Inconsistency-Based Domain Adaptation Network for Partial Transfer Intelligent Diagnosis. *IEEE Trans. Ind. Inform.* **2020**, *16*, 5965–5974. [CrossRef]
12. Ge, P.; Ren, C.X.; Dai, D.Q.; Yan, H. Domain adaptation and image classification via deep conditional adaptation network. *arXiv* **2020**, arXiv:2006.07776.
13. Li, H.; Huang, J.; Yang, X.; Luo, J.; Zhang, L.; Pang, Y. Fault Diagnosis for Rotating Machinery Using Multiscale Permutation Entropy and Convolutional Neural Networks. *Entropy* **2020**, *22*, 851. [CrossRef] [PubMed]
14. Guo, L.; Lei, Y.G.; Xing, S.; Yan, T.; Li, N. Deep convolutional transfer learning network: A new method for intelligent fault diagnosis of machines with unlabeled data. *IEEE Trans. Ind. Electron.* **2019**, *66*, 7316–7325. [CrossRef]
15. Singh, J.; Azamfar, M.; Ainapure, A.; Lee, J. Deep learning-based cross-domain adaptation for gearbox fault diagnosis under variable speed conditions. *Meas. Sci. Technol.* **2020**, *31*, 055601. [CrossRef]
16. Hasan, M.J.; Sohaib, M.; Kim, J.M. A Multitask-Aided Transfer Learning-Based Diagnostic Framework for Bearings under Inconsistent Working Conditions. *Sensors* **2020**, *20*, 7205. [CrossRef]

17. Cao, Z.; Long, M.; Wang, J.; Jordan, M. Partial Transfer Learning with Selective Adversarial Networks. In Proceedings of the IEEE Conference on Computer Vision and Pattern Recognition (CVPR), Honolulu, HI, USA, 21–26 June 2017; IEEE: Piscataway, NJ, USA, 2017; pp. 2724–2732.
18. Zhang, J.; Ding, Z.; Li, W.; Ogunbona, P. Importance Weighted Adversarial Nets for Partial Domain Adaptation. In Proceedings of the IEEE Conference on Computer Vision and Pattern Recognition (CVPR), Salt Lake City, UT, USA, 18–22 June 2018; IEEE: Piscataway, NJ, USA, 2018; pp. 8156–8164.
19. Li, X.; Zhang, W. Deep Learning-Based Partial Domain Adaptation Method on Intelligent Machinery Fault Diagnostics. *IEEE Trans. Ind. Electron.* **2020**, *99*, 1. [CrossRef]
20. Jia, S.; Wang, J.; Han, B.; Zhang, G.; Wang, X.Y.; He, J.T. A Novel Transfer Learning Method for Fault Diagnosis Using Maximum Classifier Discrepancy with Marginal Probability Distribution Adaptation. *IEEE Access* **2020**, *99*, 1. [CrossRef]
21. Busto, P.P.; Gall, J. Open Set Domain Adaptation. In Proceedings of the IEEE International Conference on Computer Vision (ICCV), Venice, Italy, 22–29 October 2017; IEEE: Piscataway, NJ, USA, 2017; pp. 754–763.
22. Saito, K.; Yamamoto, S.; Ushiku, Y.; Harada, T. Open set domain adaptation by backpropagation. In Proceedings of the European Conference on Computer Vision (ECCV), Munich, Germany, 8–14 September 2018; IEEE: Piscataway, NJ, USA, 2018; pp. 153–168.
23. You, K.; Long, M.; Cao, Z.; Wang, J.M.; Jordan, M.I. Universal Domain Adaptation. In Proceedings of the IEEE/CVF Conference on Computer Vision and Pattern Recognition (CVPR), Seattle, WA, USA, 14–19 June 2020; IEEE: Piscataway, NJ, USA, 2020; pp. 2720–2729.
24. Tao, S.; Zhang, T.; Yang, J.; Wang, X.Q.; Lu, W. Bearing fault diagnosis method based on stacked autoencoder and softmax regression. In Proceedings of the 2015 34th Chinese Control Conference (CCC), Hangzhou, China, 28–30 July 2015; IEEE: Piscataway, NJ, USA, 2015; pp. 6331–6335.
25. Saito, K.; Watanabe, K.; Ushiku, Y.; Harada, T. Maximum Classifier Discrepancy for Unsupervised Domain Adaptation. In Proceedings of the IEEE Conference on Computer Vision and Pattern Recognition (CVPR), Salt Lake City, UT, USA, 18–22 June 2018; IEEE: Piscataway, NJ, USA, 2018; pp. 3723–3732.
26. Arjovsky, M.; Chintala, S.; Bottou, L. Wasserstein GAN. *arXiv* **2017**, arXiv:1701.07875.
27. Smith, W.A.; Randall, R.B. Rolling element bearing diagnostics using the Case Western Reserve University data: A benchmark study. *Mech. Syst. Signal Process.* **2015**, *64*, 100–131. [CrossRef]
28. Li, X.; Zhang, W.; Ma, H.; Luo, Z.; Li, X. Partial transfer learning in machinery cross-domain fault diagnostics using class-weighted adversarial networks. *Neural Netw.* **2020**, *129*, 313–322. [CrossRef] [PubMed]
29. Han, B.; Zhang, G.; Wang, J.; Wang, X.Y.; Jia, S.X.; He, J.T. Research and application of regularized sparse filtering model for intelligent fault diagnosis under large speed fluctuation. *IEEE Access* **2020**, *8*, 39809–39818. [CrossRef]
30. Wang, J.; Li, S.; An, Z.; Jiang, X.X.; Qian, W.W.; Ji, S.S. Batch-normalized deep neural networks for achieving fast intelligent fault diagnosis of machines. *Neurocomputing* **2019**, *15*, 53–65. [CrossRef]
31. Laurens, V.D.M.; Geoffrey, H. Visualizing Data using t-SNE. *J. Mach. Learn. Res.* **2008**, *9*, 2579–2605.

Article

Low-Pass Filtering Empirical Wavelet Transform Machine Learning Based Fault Diagnosis for Combined Fault of Wind Turbines

Yancai Xiao *, Jinyu Xue, Mengdi Li and Wei Yang

School of Mechanical, Electronic and Control Engineering, Beijing Jiaotong University, Beijing 100044, China; 20126082@bjtu.edu.cn (J.X.); 18121281@bjtu.edu.cn (M.L.); 19116025@bjtu.edu.cn (W.Y.)
* Correspondence: ycxiao@bjtu.edu.cn

Abstract: Fault diagnosis of wind turbines is of great importance to reduce operating and maintenance costs of wind farms. At present, most wind turbine fault diagnosis methods are focused on single faults, and the methods for combined faults usually depend on inefficient manual analysis. Filling the gap, this paper proposes a low-pass filtering empirical wavelet transform (LPFEWT) machine learning based fault diagnosis method for combined fault of wind turbines, which can identify the fault type of wind turbines simply and efficiently without human experience and with low computation costs. In this method, low-pass filtering empirical wavelet transform is proposed to extract fault features from vibration signals, LPFEWT energies are selected to be the inputs of the fault diagnosis model, a grey wolf optimizer hyperparameter tuned support vector machine (SVM) is employed for fault diagnosis. The method is verified on a wind turbine test rig that can simulate shaft misalignment and broken gear tooth faulty conditions. Compared with other models, the proposed model has superiority for this classification problem.

Keywords: combined fault diagnosis; empirical wavelet transform; grey wolf optimizer; low pass FIR filter; support vector machine

Citation: Xiao, Y.; Xue, J.; Li, M.; Yang, W. Low-Pass Filtering Empirical Wavelet Transform Machine Learning Based Fault Diagnosis for Combined Fault of Wind Turbines. *Entropy* 2021, 23, 975. https://doi.org/10.3390/e23080975

Academic Editor: Sotiris Kotsiantis

Received: 24 June 2021
Accepted: 26 July 2021
Published: 29 July 2021

Publisher's Note: MDPI stays neutral with regard to jurisdictional claims in published maps and institutional affiliations.

Copyright: © 2021 by the authors. Licensee MDPI, Basel, Switzerland. This article is an open access article distributed under the terms and conditions of the Creative Commons Attribution (CC BY) license (https://creativecommons.org/licenses/by/4.0/).

1. Introduction

With the improvement of people's environmental awareness, sustainable and carbon-neutral renewable energy has gradually developed to replace oil, coal and other traditional fossil fuels [1]. According to a recent report about renewable capacity statistics [2], the world's wind energy capacity is 622,704 MW in 2019, accounting for 24.55% of the total renewable energy capacity, second only to the hydropower which is the oldest renewable energy source [3]. The annual growth rate of wind energy is 10.44% in 2019, second only to the rapidly developing solar energy. Improving the efficiency of wind turbines has always been a hot issue in terms of wind energy utilization. In addition to study the selection of wind turbine [4–6], it is useful to reasonably design the wind turbines' structure [7,8]. At the same time, wind turbines are usually exposed to dynamic and harsh weather conditions, experiencing variable and rough working environments, which makes them prone to failure than other ordinary machinery. If a component of the wind turbine is broken without awareness of workers, it may well cause damage to other components, and even lead to the shutdown of the wind turbine, resulting in huge economic losses [9]. Operating and maintenance costs account for more than 25% of total costs for onshore wind farms and these costs are even higher for offshore projects [10]. Therefore, it is of great significance to reduce maintenance costs and improve the efficiency of wind farms by detecting the fault of wind turbines in time.

Many studies have been carried out on fault diagnosis of wind turbines. Such as Liu et al. [11] introduced local mean decomposition (LMD) to analyze the wind turbine gearbox vibration signals for fault diagnosis. Feng et al. [12] proposed a frequency demodulation analysis method based on the ensemble empirical mode decomposition (EEMD) and energy separation algorithm to detect and locate the fault of wind turbine planetary gearbox by analyzing vibration signals. Chen et al. [13] applied empirical wavelet transformation (EWT) to vibration signals to diagnose wind turbine generator bearings faults. Those methods depend on experienced people to analyze the signal and determine the fault of drivetrains of wind turbines, although the precision is guaranteed, it is lack of efficiency. In recent years, with the rise of machine learning (ML), some scholars have tried to use ML methods to diagnosis the drivetrain of wind turbines. For example, Liu et al. [14] extracted features from vibration signals by diagonal spectrum and employed clustering binary tree support vector machines to diagnosis the wind turbines gearbox. Tang et al. [15] proposed a fault diagnosis method for the drivetrain of wind turbines based on manifold learning and Shannon wavelet support vector machine. Gao et al. [16] decomposed vibration signals by integral extension local mean decomposition (IELMD) and calculated multiscale entropy values as features for least squares support vector machines to identify fault type of rolling bearing in wind turbine gearbox. Lei et al. [17] introduced long-short term memory (LSTM) networks in wind turbine fault diagnosis. Jiang et al. [18] proposed multiscale convolutional neural network (MSCNN) to diagnose wind turbine gearbox faults.

Almost two-thirds of ML-based wind turbine fault diagnosis methods use classification, whose procedures include preprocess data, equalize classes, feature extraction, feature selection, hyperparameter tuning, cross-validation and use the best model [19]. This intelligent way allows the diagnosis to be free from expert experience.

However, most of these ML-based wind turbine fault diagnosis methods only studied on single fault [15–19]. In reality, a wind turbine is a complex system, failures could happen one after another or simultaneously, therefore, a wind turbine may have more than one fault at the same time, i.e., combined fault occurs. For example, misalignment may lead to gear or bearing fails, then multiple faults coexist. Gear faults in different stages is also a common combined fault [20]. Combined fault (also called compound fault) is more difficult to diagnose than single fault because typical fault features will become difficult to be extracted. At present, combined fault diagnosis of wind turbines usually depends on manual analysis to calculate, extract and show the frequencies of different faults in spectrums [21–27]. Only a few scholars have studied combined fault diagnosis by ML. For example, Zhong et al. [28] decomposed the vibration signal into a series of intrinsic mode functions (IMFs) by Hilbert-Huang transform (HHT) with ensemble empirical mode decomposition (EEMD), then selected useful IMFs by correlation coefficients, and calculated the energy vector from the selected IMFs together with maximum amplitude and corresponding frequency and six time-domain statistical indices as features of pairwise-coupled sparse Bayes extreme learning machine to detect several common gearbox single-faults and simultaneous-faults.

This paper will focus on a ML-based fault diagnosis method for combined faults and single faults of wind turbines. In our method, a composite fault is considered as a fault equivalent to a single fault, which means the output of a combined fault is not multiple binary tags for each single fault (multilabel classification problem). The reminder of this paper is structured as follows: Section 2 introduces the proposed method and related theories. Section 3 presents the test rig, the experiments and the results. Finally, the conclusion in Section 4.

2. Methods

The fault diagnosis method for combined fault of wind turbines we proposed can be described as follows. First, extract features from vibration signals by low pass filtering empirical wavelet transform (LPFEWT). Then, build features datasets in different conditions (normal, single faults and combined fault). Last, train the support vector machine (SVM) for classification, using grey wolf optimizer (GWO) for hyperparameter tuning. After training, the obtained SVM model can identify faults of wind turbines by inputting features of vibration signals. The flow chart of the method is shown in Figure 1.

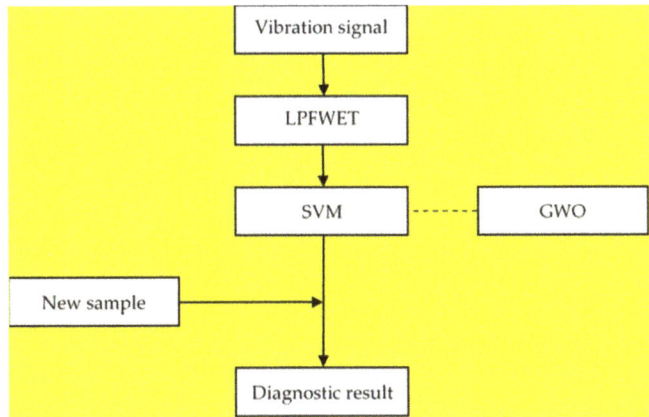

Figure 1. The flow chart of the proposed ML-based fault diagnosis method for combined fault of wind turbines.

2.1. Low Pass Filtering Empirical Wavelet Transform (LPFEWT)

Empirical Wavelet Transform (EWT) is a new adaptive signal processing approach proposed by Gilles in 2013 [29]. The main idea is to adaptively decompose the modes of a signal from its Fourier spectrum by an appropriately built wavelet filter bank. The steps of EWT are summarized as follows:

- Fast Fourier Transform (FFT);

Convert the signal f to the frequency domain by FFT to get its Fourier spectrum (frequency $\omega \in [0, \pi]$).

- Fourier Spectrum Segmentation;

Divide the Fourier spectrum into N contiguous segments. Let ω_n denote the limits between each segment. Each segment is denoted as $\Lambda_n = [\omega_{n-1}, \omega_n]$. With each ω_n as center, a transition phase of width $2\tau_n$ is defined.

- Mode Extraction;

Let $\hat{f}$ and $\check{f}$ denote the Fourier transform and its inverse respectively. Choose τ_n proportional to ω_n: $\tau_n = \gamma \omega_n$, where $0 < \gamma < 1$. Consequently, $\forall n > 0$, the empirical scaling function $\hat{\phi}_n(\omega)$ and the empirical wavelets $\hat{\psi}_n(\omega)$ are as follows:

$$\hat{\phi}_n(\omega) = \begin{cases} 1, & |\omega| \leq (1-\gamma)\omega_n \\ \cos\left[\frac{\pi}{2}\beta\left(\frac{1}{2\tau_n}(|\omega| - (1-\gamma)\omega_n)\right)\right], \\ \quad (1-\gamma)\omega_n \leq |\omega| \leq (1+\gamma)\omega_n \\ 0, & otherwise \end{cases} \quad (1)$$

and

$$\hat{\psi}_n(\omega) = \begin{cases} 1, & \omega_n + \tau_n \leq |\omega| \leq (1-\gamma)\omega_{n+1} \\ \cos\left[\frac{\pi}{2}\beta\left(\frac{1}{2\tau\omega_{n+1}}(|\omega| - (1-\gamma)\omega_{n+1})\right)\right], \\ \quad (1-\gamma)\omega_{n+1} \leq |\omega| \leq (1+\gamma)\omega_{n+1} \\ \sin\left[\frac{\pi}{2}\beta\left(\frac{1}{2\gamma\omega_n}(|\omega| - (1-\gamma)\omega_n)\right)\right], \\ \quad (1-\gamma)\omega_n \leq |\omega| \leq (1+\gamma)\omega_n \\ 0, & otherwise \end{cases} \quad (2)$$

To construct a tight frame set of empirical wavelets, choose

$$\gamma < min_n \left(\frac{\omega_{n+1} - \omega_n}{\omega_{n+1} + \omega_n} \right) \tag{3}$$

The detail coefficients $W_f^\varepsilon(n,t)$ are given by the inner products with the empirical wavelets function $\hat{\psi}_n(\omega)$, and the approximation coefficients $W_f^\varepsilon(0,t)$ are given by the inner product with the scaling function $\hat{\phi}_1(\omega)$.

$$\begin{aligned} W_f^\varepsilon(n,t) &= \langle f, \psi_n \rangle = \int f(\tau) \overline{\psi_n(\tau - t)} d\tau \\ &= \left(\hat{f}(\omega) \overline{\hat{\psi}_n(\omega)} \right) \end{aligned} \tag{4}$$

$$\begin{aligned} W_f^\varepsilon(0,t) &= \langle f, \phi_1 \rangle = \int f(\tau) \overline{\phi_1(\tau - t)} d\tau \\ &= \left(\hat{f}(\omega) \overline{\hat{\phi}_1(\omega)} \right) \end{aligned} \tag{5}$$

The reconstruction is obtained by

$$\begin{aligned} f(t) &= W_f^\varepsilon(0,t) \star \phi_1(t) + \sum_{n=1}^N W_f^\varepsilon(n,t) \star \psi_n(t) \\ &= \left(\hat{W}_f^\varepsilon(0,\omega) \hat{\phi}_1(\omega) + \sum_{n=1}^N \hat{W}_f^\varepsilon(n,t) \hat{\psi}_n(t) \right) \end{aligned} \tag{6}$$

There are multiple algorithms to automatically segment the Fourier spectrum, such as local-maxima, local-maxima-minima and scale-space (including otsu, half-normal, empirical law, means and k-means) [29,30]. The scale-space algorithms are parameterless, but it takes long time for the computation when processing a long signal. And different signals are often decomposed into different amounts of modes, which is inconvenient for the comparison with each other. Considering these factors, we choose the simplest and fastest algorithm–local-maxima, which can set the max number of segments.

Based on EWT, LPFEWT is proposed to extract features. First, design a low pass FIR filter with an appropriate cut-off frequency for the signal. Next, employ EWT on the filtered signal to decompose the signal into several empirical modes. Then, exclude the empirical mode of the highest frequencies which is mostly affected by the filter. Last, calculate the indices of the left modes as features. According to this approach, the feature required for fault diagnosis can be obtained easily.

Compared to the tradition wavelet transform, LPFEWT is adaptive, which means it decomposes the signal based on the information contained in the signal itself so that there is no need to choose or design specific wavelet basis for the signal.

2.2. Support Vector Machine (SVM)

SVM is a very powerful and versatile ML model and particularly well suited for classification of complex but small- or medium-sized datasets [31].

The simplest linear SVM for binary classification can be described as follows. For all samples to be classified $x_i (i = 1, 2, \ldots, m)$, the output is

$$y_i = sign\left(\boldsymbol{w}^T \boldsymbol{x}_i + b \right) \tag{7}$$

i.e., $y_i = -1$ if $\boldsymbol{w}^T \boldsymbol{x}_i + b < 0$, $y_i = +1$ if $\boldsymbol{w}^T \boldsymbol{x}_i + b > 0$. So the hyperplane $\boldsymbol{w}^T \boldsymbol{x} + b = 0$ is decision boundary. To make the decision boundary best for separation, construct two hyperplanes $\boldsymbol{w}^T \boldsymbol{x} + b = -1$ and $\boldsymbol{w}^T \boldsymbol{x} + b = 1$ which are parallel and at equal distance to the decision boundary, i.e., $y_i = -1$ if $\boldsymbol{w}^T \boldsymbol{x}_i + b \leq -1$, $y_i = +1$ if $\boldsymbol{w}^T \boldsymbol{x}_i + b \geq 1$. Training

SVM means finding the value of w and b that make the width of the margin $2/\|w\|$ as large as possible. That is a constrained optimization problem

$$\max_{w,b} \frac{2}{\|w\|} \quad s.t. \quad y_i(w^T x_i + b) \geq 1, \ i = 1, 2, \ldots, n \tag{8}$$

which can be converted to an equivalent problem

$$\min_{w,b} \frac{1}{2}\|w\|^2 \quad s.t. \quad y_i(w^T x_i + b) \geq 1, \ i = 1, 2, \ldots, n \tag{9}$$

This is a convex quadratic optimization problem with linear constraints, which is known as quadratic programming (QP) problems and can be solved by the method of Lagrange multipliers. Introduce Lagrange multipliers $\lambda = (\lambda_1, \lambda_2, \cdots, \lambda_m)$, the objective function of optimization can be expressed as

$$\mathcal{L}(w,b,\lambda) = \frac{1}{2}\|w\|^2 + \sum_{i=1}^{n} \lambda_i \left[1 - y_i(w^T x_i + b)\right] \tag{10}$$
$$\lambda_i \geq 0, i = 1, 2, \cdots, n$$

The problem is to solve

$$\min_{w,b} \max_{\lambda} \mathcal{L}(w,b,\lambda) \tag{11}$$

The dual problem is

$$\max_{\lambda} \min_{w,b} \mathcal{L}(w,b,\lambda) \tag{12}$$

Calculate the gradients of both w and b, and set them equal to zero.

$$\nabla_w \mathcal{L}(w,b,\lambda) = w - \sum_{i=1}^{n} \lambda_i x_i y_i = 0 \tag{13}$$

$$\frac{\partial}{\partial b}\mathcal{L}(w,b,\lambda) = -\sum_{i=1}^{n} \lambda_i y_i = 0 \tag{14}$$

Substitute (13) and (14) into problem (12), obtain

$$\max_{\lambda} \sum_{i=1}^{n} \lambda_i - \frac{1}{2} \sum_{i=1}^{n} \sum_{j=1}^{n} \lambda_i \lambda_j y_i y_j (x_i \cdot x_j) \tag{15}$$
$$s.t. \sum_{i=1}^{n} \lambda_i y_i = 0, \lambda_i \geq 0, i = 1, 2, \ldots, m$$

Consequently, the original minimization problem about w and b is converted to a QP problem about solving λ.

To make the model more flexible, soft margin classification is proposed which allows few instances between the margins or even on the wrong side. Soft margin SVM introduces slack variable $\xi_i (i = 1, 2, \cdots, n)$, so the problem becomes

$$\min_{w,b,\xi} \frac{1}{2}\|w\|^2 + C \sum_{t=1}^{n} \xi_i \tag{16}$$
$$s.t. \quad y_i(w^T x_i + b) \geq 1 - \xi_i, \ \xi_i > 0, \ i = 1, 2, \ldots, m$$

where C is penalty term. The bigger the C, the more penalty SVM gets when it makes misclassification, the less the tolerance, the smaller the margin.

The QP problem equivalent to soft margin SVM classification is

$$\max_{\lambda} \sum_{i=1}^{n} \lambda_i - \frac{1}{2} \sum_{i=1}^{n} \sum_{j=1}^{n} \lambda_i \lambda_j y_i y_j (x_i \cdot x_j) \quad (17)$$
$$s.t. \sum_{i=1}^{n} \lambda_i y_i = 0, 0 \leq \lambda_i \leq C, i = 1, 2, \ldots, m$$

For problems that are not linearly separable, transformation ϕ is introduced to map x from the original space to a higher dimensional space $\phi(x)$, which makes it easier to find a linear decision boundary in the new feature space. The kernel function $K(x_i, x_j) = \phi(x_i) \cdot \phi(x_j)$ is proposed to focus on the results without computing the coordinates of the data in the new space. The kernel trick makes the whole process much more computationally efficient. Problem (17) can be rewritten as

$$\max_{\lambda} \sum_{i=1}^{n} \lambda_i - \frac{1}{2} \sum_{i=1}^{n} \sum_{j=1}^{n} \lambda_i \lambda_j y_i y_j K(x_i, x_j) \quad (18)$$
$$s.t. \sum_{i=1}^{n} \lambda_i y_i = 0, 0 \leq \lambda_i \leq C, i = 1, 2, \ldots, m$$

In this paper, we use radial basis function (RBF) kernel as below

$$K(x_i, x_j) = e^{-\gamma \|x_i - x_j\|^2}, \gamma > 0 \quad (19)$$

RBF kernel is one of the most used kernel functions, which can deal with both linear and nonlinear classification problems. The result of linear classification using RBF kernel is comparable to using linear kernel [32,33].

2.3. Grey Wolf Optimizer

Grey Wolf Optimizer (GWO) is a swarm intelligence (SI) algorithm proposed by Mirjalili et al. [34] in 2014 that imitates the leadership hierarchy and hunting mechanism of grey wolves in nature. In this paper, it is used to optimize the parameters in SVM. The social hierarchy of gray wolves is shown in Figure 2. Grey wolves are divided into four levels from α to ω. The upper level wolves dominate the lower level ones, and the lower level wolves follow the upper level ones.

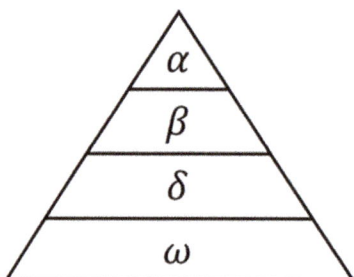

Figure 2. The social hierarchy of grey wolves.

In the GWO algorithm, imitating the social hierarchy of grey wolves, the first best candidate solution is regarded as α, the second best candidate solution is regarded as β, the third best candidate solution is regarded as δ, the remaining candidate solutions are regarded as ω. The hunting (optimization) is guided by α, β and δ, while ω follow them. The encircling behavior is modeled as follows:

$$D = |C \cdot X_p(t) - X(t)| \quad (20)$$

$$X(t+1) = X_p(t) - A \cdot D \tag{21}$$

where t represents the number of iterations, A and C are coefficient vectors, X_p is the position vector of the prey (optimum), X is the position vector of a grey wolf, and D represents the distance between the grey wolf and the prey.

The vectors A and C are defined as follows:

$$A = 2a \cdot r_1 - a \tag{22}$$

$$C = 2 \cdot r_2 \tag{23}$$

where components of a are linearly dropped from 2 to 0 over the course of iterations, components of r_1 and r_2 are random numbers in $[0, 1]$.

The random vectors r_1 and r_2 allow grey wolves to move any position within a certain range of the prey. With the vector a decreases, grey wolves encircle and pursue the prey. The location of the prey is replaced by the decisions of all three grey wolves α, β and δ. The following equations are used for updating the position of each grey wolf.

$$\begin{cases} D_\alpha = |C_1 \cdot X_\alpha - X(t)| \\ D_\beta = |C_2 \cdot X_\beta - X(t)| \\ D_\delta = |C_3 \cdot X_\delta - X(t)| \end{cases} \tag{24}$$

$$\begin{cases} X_1 = X_\alpha - A_1 \cdot D_\alpha \\ X_2 = X_\beta - A_2 \cdot D_\beta \\ X_3 = X_\delta - A_3 \cdot D_\delta \end{cases} \tag{25}$$

$$X(t+1) = \frac{X_1 + X_2 + X_3}{3} \tag{26}$$

Since A is a random vector in the interval $[-a, a]$, the next position of wolves will approach the prey if $|A| < 1$, and move away from the prey if $|A| > 1$. This means that grey wolves not only pursue and attack current prey but also leave to search for other prey. In other words, the GWO algorithm has exploration feature to help avoid local optima. The random vector C simulates the obstacles to approaching prey in nature.

GWO can make the process of hyperparameter tuning of SVM more effective than normal way (grid search or randomized search). Also, GWO hyperparameter tuned has better classification accuracy than the typical one-versus-one multi-class SVM [35]. Compared with particle swarm optimization (PSO), GWO has fewer parameters to be determined, only the population and the max number of iterations, because it updates the positions of search agents by the positions of the three best wolves, while PSO updates the positions of search agents by the global best position and the personal best position, and each search agent has velocity besides position.

3. Experimental Results

3.1. Experimental Test Rig and Data Collection

The laboratory's wind turbine drivetrain fault test rig is shown in Figure 3, which consists of a control panel cabinet and an experimental test bench to simulate doubly-fed induction generator (DFIG) wind turbine shaft misalignment (between the gearbox and the generator) and broken gear tooth faulty conditions. In Figure 3a, the speed of the motor of the experimental test bench on the right side is decelerated by a planetary gear reducer to simulate the wind blowing blade speed, then it is accelerated by a planetary gear accelerator and a gearbox to drive the generator. The maximum speed of the driving motor is 720 r/min, the speed of the generator is 500 r/min. The left gearbox can be adjusted by the handle to select a normal gear or a broken gear. The generator can be adjusted by the support to create offset or angular misalignment. The control panel cabinet shown in Figure 3b can set and display the motor speed, showing the angle between the generator and the gearbox and other electrical parameters.

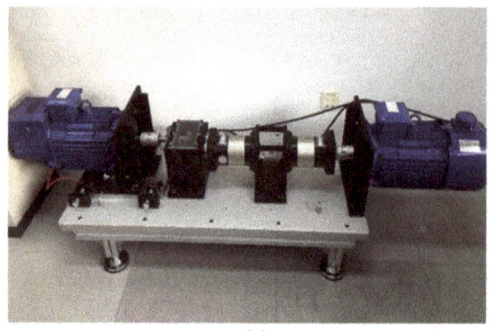
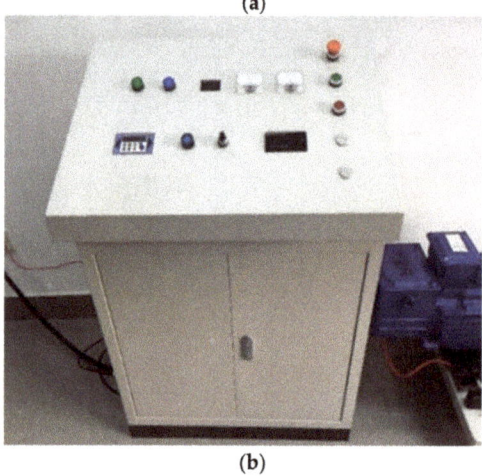

Figure 3. Wind turbine drivetrain fault experimental test rig: (**a**) experimental test bench; (**b**) control panel cabine.

The vibration signals in normal, misalignment, broken tooth and combined fault (misalignment and tooth broken) conditions were collected from the test rig. Set two measuring point, at the vertical and horizontal direction of the gearbox high-speed output shaft side, with a sampling frequency of 1 kHz and a sample time of 20 s. In the normal and broken tooth conditions, 18 sets of data were collected at the motor speed from 200 r/min to 720 r/min respectively. In misalignment condition, 26 sets of data were collected at the motor speed from 200 r/min to 680 r/min. In combined fault condition, 10 sets of data were collected at the motor speed from 200 r/min to 520 r/min. After preliminary frequency domain analysis of the signals, only the vertical direction signal is used for diagnosis in this paper. With non-overlapping 10,000 points of the signal, the samples in different conditions are shown in Figure 4, from which it can be seen that the presence of broken tooth is easy to distinguish, while the presence of misalignment is not.

3.2. LPFEWT and Comparison with Other Approaches

Employ LPFEWT to extract features from the signal. The cut-off frequency of the low-pass filter is 50 Hz, about 6 times the rated rotating frequency of the generator. The magnitude and phase responses of the designed 40th-order Hamming Window FIR low-pass filter are shown in Figure 5. The filtered signal is decomposed by EWT and the number of EWT Fourier spectrum segments is set to 6. The EWT decomposition results of a combined fault signal are shown in Figure 6, obtained 6 empirical mode components from low frequency to high frequency. Discard the highest frequency component (the 6th mode) and calculate features of the left 5 empirical modes.

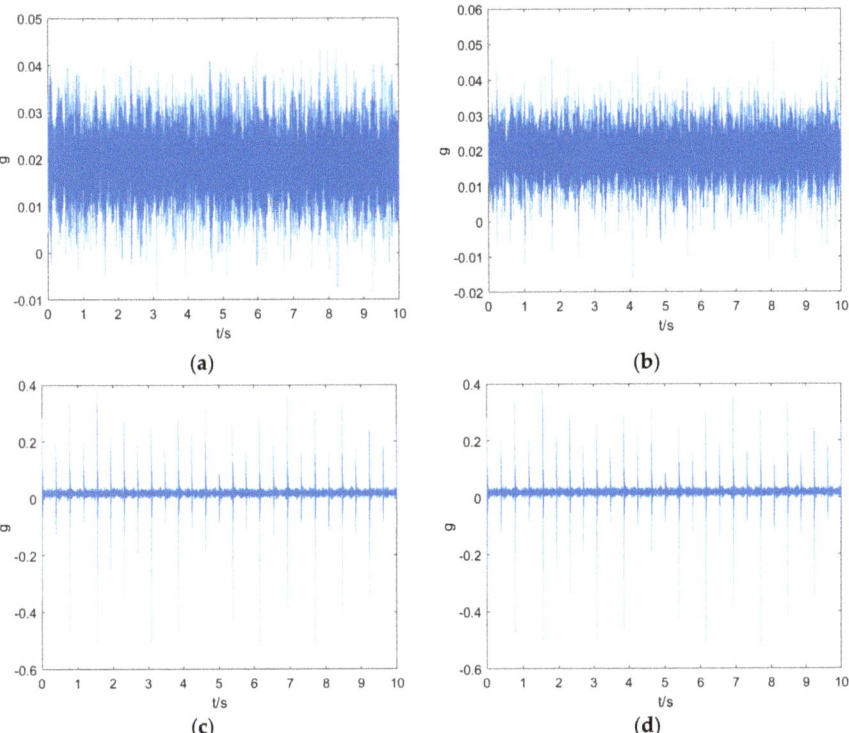

Figure 4. Samples of vibration signals in different conditions: (**a**) normal; (**b**) misalignment; (**c**) broken tooth; (**d**) combined fault.

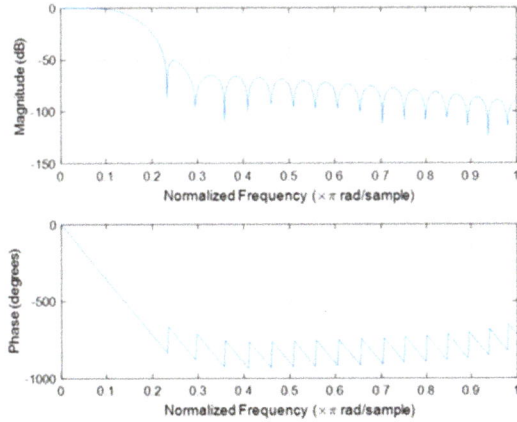

Figure 5. The magnitude and phase responses of the designed FIR low-pass filter.

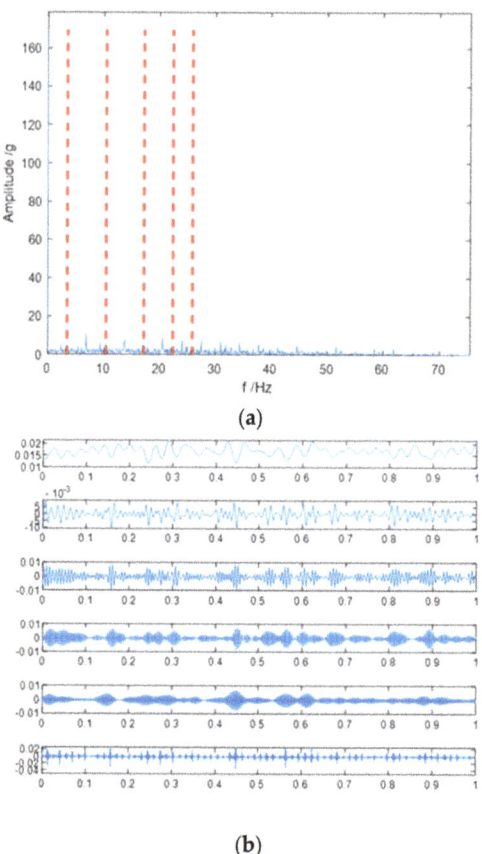

Figure 6. The EWT decomposition results of a combined fault signal: (**a**) Fourier spectrum segmentation; (**b**) empirical mode components.

We choose energies of the components as features, that is, the sum of the squares of the amplitude. There are 20 combined fault samples, 27 broken tooth samples, 26 misalignment samples and 27 normal samples, 100 samples in total. Shuffle the dataset and save. Take 14 combined fault samples, 18 broken tooth samples, 18 misalignment samples and 18 normal as training set. The remaining 32 samples of the dataset is testing set. We use LIBSVM Version 3.24 package for SVM classification under MATLAB 2018b. Train the SVM classification model for fault diagnosis, using GWO algorithm search the optimum values of penalty term C and RBF kernel parameter γ in the range of $[0.01, 100]$. The average accuracy of 3-fold cross-validation of the training set is used as the fitness of the agents. The grey wolf population is set as 100 and the iteration is set as 50. Empirical modal decomposition (EMD) which is similar to EWT is chosen for comparison. Energies of components obtained by different approaches are inputs of the SVM model. Figure 7 shows the confusion matrix obtained by inputting the components energies of different methods into the SVM model. The horizontal direction represents the predicted class, and the vertical direction represents the true class. The 4×4 matrix is the number of samples of each type, and the percentage includes the prediction accuracy rate, false alarm rate and missing alarm rate of each type. Comparison of results are shown in Table 1. Different approaches with 'LPF' prefix use the same FIR low-pass. All approaches use same amounts of components of the signals.

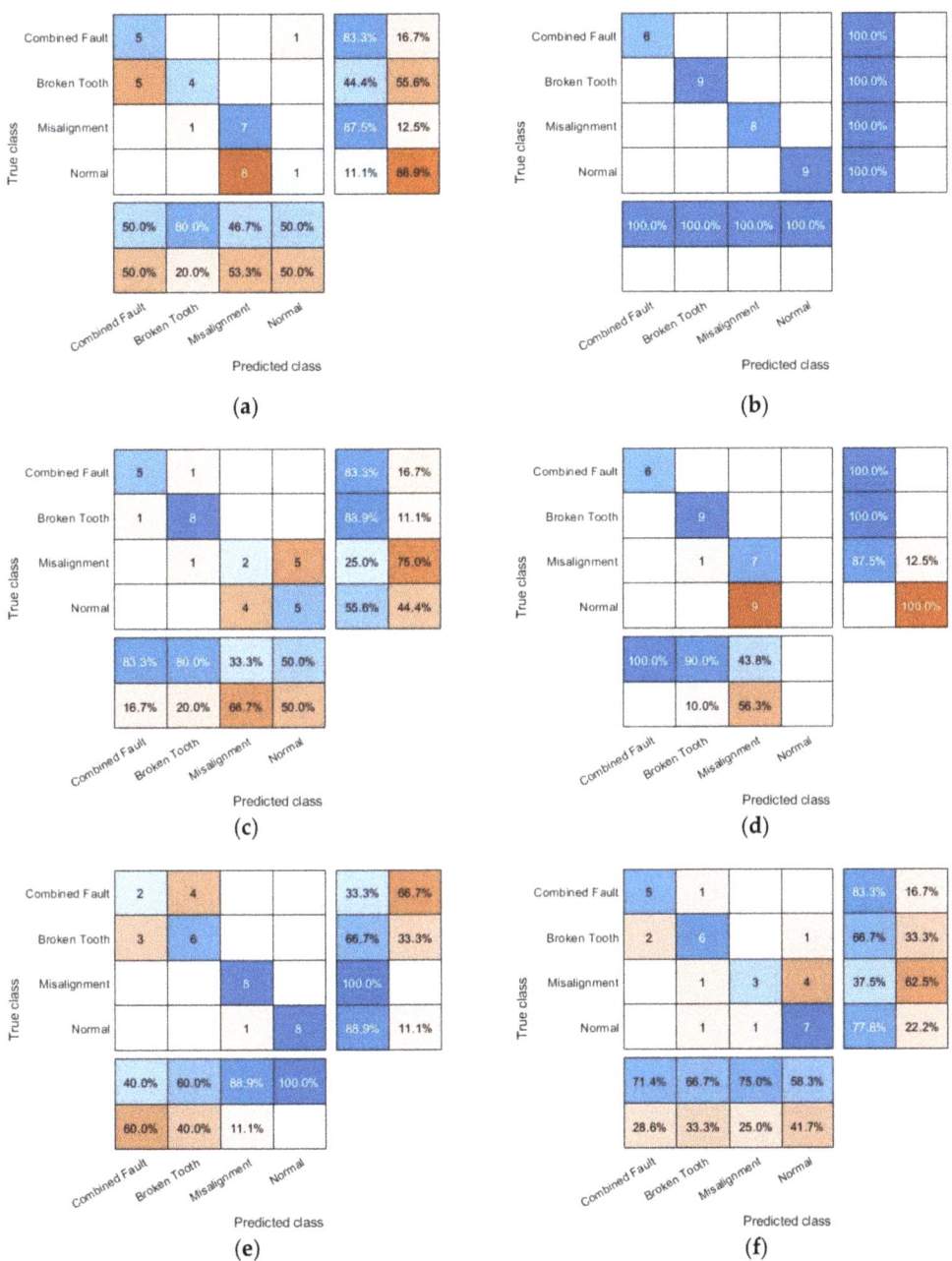

Figure 7. Confusion matrix charts of fault diagnosis results with features obtained by different approaches: (**a**) EWT; (**b**) LPFEWT; (**c**) EMD high frequency components; (**d**) LPFEMD high frequency components; (**e**) EMD low frequency components; (**f**) LPFEMD low frequency components.

Table 1. Comparison of Fault Diagnosis Results with Different Feature Extraction Approaches.

Approach	C	γ	Training Set Accuracy	Testing Set Accuracy	False Alarm Rate	Missing Alarm Rate
EWT	98.135258	4.997962	80.8824% (55/68)	53.125% (17/32)	88.9% (8/9)	4.3% (1/23)
LPFEWT	66.953529	57.624745	94.1176% (64/68)	100% (32/32)	0% (0/9)	0% (0/23)
EMD high frequency components	17.297601	39.468164	76.4706% (52/68)	68.75% (22/32)	44.4% (4/9)	21.7% (5/23)
LPFEMD high frequency components	45.388002	96.255492	76.4706% (52/68)	62.5% (20/32)	100% (9/9)	0% (0/23)
EMD low frequency components	26.988942	37.129502	85.2941% (58/68)	75% (24/32)	11.1% (1/9)	0% (0/23)
LPFEMD low frequency components	48.145791	1.052425	69.1176% (47/68)	65.625% (21/32)	22.2% (2/9)	21.7% (5/23)

From Figure 7 and Table 1 we can see, the testing set accuracy of using EWT directly is low, only 53.125%, and there is a lot of conditions confusions. Using LPFEWT to extract time-frequency domain features, the testing set accuracy is highly improved, reaching 100%. In addition, using EWT directly has high false alarm rate, while LPFEWT solves this problem. Among approaches based on EMD, EMD low frequency components has the highest accuracy and the lowest false alarm rate and missing alarm rate, which is 75%, but there are confusions between combined fault and broken tooth or misalignment and normal condition. LPFEMD low frequency components can only identify combined fault and broken tooth correctly. Both with or without the low-pass filter, EMD low frequency components has lower false alarm rate than the high frequency components. Both using high and low frequency components, the accuracy of LPFEMD is lower than that of EMD, and the false alarm rate is higher. The use of low-pass filter in diagnosis with approaches based on EMD will decrease the accuracy instead of increase that, and increase the false alarm rate. Among the six approaches of feature extraction, LPFEWT has the best performance.

We also tried SVM with linear kernel, the accuracy of training set and testing set are 82.4% and 87.5% respectively. So the classification of the dataset is a nonlinear problem, using RBF kernel is proper.

3.3. LPFEWT with Different Number of Fourier Spectrum Segments

To explore the effect of the number of LPFEWT Fourier spectrum segments on fault diagnosis results, the diagnosis was carried out with different number of Fourier spectrum segments, using energies of empirical modes as features, the results are shown in Table 2.

Table 2. Diagnosis Results of Employing LPFEWT with Different Number of Fourier Spectrum Segments.

Number of Segments	C	γ	Training Set Accuracy	Testing Set Accuracy
3	54.450584	44.708328	88.2353% (60/68)	100% (32/32)
4	43.410799	96.515668	92.6471% (63/68)	100% (32/32)
5	49.290038	78.087215	94.1176% (64/68)	100% (32/32)
6	66.953529	57.624745	94.1176% (64/68)	100% (32/32)
7	60.868225	95.642439	94.1176% (64/68)	100% (32/32)
8	98.149020	74.985752	94.1176% (64/68)	100% (32/32)
9	80.564115	91.484842	95.5882% (65/68)	96.875% (31/32)

From Table 2, it can be seen that when the number of LPFEWT Fourier spectrum segments is small, although the testing set has good accuracy, the training set accuracy is slightly lower. When the number of LPFEWT Fourier spectrum segments is 5, 6, 7, 8, the diagnosis performance does not change. When the number of LPFEWT Fourier spectrum segments is 9, the accuracy of training set is improved a little, but the accuracy of testing set is reduced. Therefore, the number of LPFEWT Fourier spectrum segments should not be too small or too large, and there is a range of proper number of segments. It is suggested that the number of LPFEWT Fourier spectrum segments is set to 6 first, if the diagnosis results is not good enough, increase the number of segments one by one.

3.4. Effectiveness of the Proposed SVM Based Method

In the proposed method, we choose SVM for classification because it has superiority when dealing with small datasets. Since the samples of wind turbines in faults are relatively few. So deep learning which needs a large dataset is not suitable. Considering the speed of prediction after training, k-nearest neighbors (k-NN) algorithm which computes the distances between the instance and all the training instances to make decisions is abandoned. We compared SVM with naive Bayes, decision trees, random forests and artificial neural networks (ANN), the results are shown in Table 3.

Table 3. Comparison Results of Different ML Classification Models.

Model	Training Set Accuracy	Testing Set Accuracy	False Alarm Rate	Missing Alarm Rate
SVM	94.1176% (64/68)	100% (32/32)	0% (0/9)	0% (0/23)
Naive Bayes	95.5882% (65/68)	96.875% (31/32)	0% (0/9)	0% (0/23)
Decision trees	89.7059% (61/68)	100% (32/32)	0% (0/9)	0% (0/23)
Random forests	97.0588% (66/68)	96.875% (31/32)	0% (0/9)	0% (0/23)
ANN	92.6471% (63/68)	96.875% (31/32)	0% (0/9)	0% (0/23)

From Table 3, we can see, for this classification problem, SVM has the best training performance and the accuracy of the training set is 94.1176%. The decision trees model has the lowest accuracy on training set with the highest accuracy as SVM model on testing set. All the models have good generalization ability. This show the feature selected is powerful. SVM has the best testing set accuracy and medium training set accuracy. Obviously, SVM is the best choice for this particular wind turbine fault diagnosis problem, which has good generalization ability even on a small dataset and easy to use (only has two hyperparameters need to tune).

4. Conclusions

This paper studies a ML-based fault diagnosis method for combined fault of wind turbines. LPFEWT is proposed to extract time-frequency domain features from vibration signals. And a GWO hyperparameter tuned SVM is employed for fault diagnosis. The method is verified on a DFIG wind turbine drivetrain fault test rig in the laboratory. The experimental results show that LPFEWT can greatly improve the accuracy of fault diagnosis and it is superior to other feature extraction approaches. The effect of the number of LPFEWT Fourier spectrum segments on fault diagnosis results is explored and a reasonable strategy to choose the number of segments is given. SVM is proved to be superior in this classification problem.

Compared with the existing analysis methods for combined fault, this ML-based method is efficient. After training the ML model at low computation costs, it can quickly handle the data of wind turbines working at different speeds and easily identify the faults without human knowledge. The method can also be applied to fault diagnosis of other rotating machinery.

Author Contributions: Conceptualization, Y.X. and M.L.; methodology, Y.X. and M.L.; software, M.L.; validation, Y.X. and J.X.; resources, Y.X.; data curation, M.L. and W.Y.; writing—original draft preparation, M.L.; writing—review and editing, Y.X. and J.X.; supervision, Y.X.; project administration, Y.X. All authors have read and agreed to the published version of the manuscript.

Funding: This research was funded by the National Natural Science Foundation of China (51577008).

Conflicts of Interest: The authors declare no conflict of interest.

References

1. Rehman, S.; Khan, S.A.; Alhems, L.M. A review of wind-turbine structural stability, failure and alleviation. *Wind. Struct.* **2020**, *30*, 511–524.
2. IRENA. *Renewable Capacity Statistics 2020*; International Renewable Energy Agency (IRENA): Abu Dhabi, United Arab Emirates, 2020.
3. Mahmud, M.A.P.; Huda, N.; Farjana, S.H.; Lang, C. Environmental sustainability assessment of hydropower plant in Europe using life cycle assessment. *IOP Conf. Ser. Mater. Sci. Eng.* **2018**, *351*, 1–8. [CrossRef]
4. Rehman, S.; Khan, S.A.; Alhems, L.M. A Rule-Based Fuzzy Logic Methodology for Multi-Criteria Selection of Wind Turbines. *Sustainability* **2020**, *12*, 8467. [CrossRef]
5. Rehman, S.; Khan, S.A. Fuzzy Logic Based Multi-Criteria Wind Turbine Selection Strategy—A Case Study of Qassim, Saudi Arabia. *Energies* **2016**, *9*, 872. [CrossRef]
6. Rehman, S.; Khan, S.A.; Alhems, L.M. Application of TOPSIS Approach to Multi-Criteria Selection of Wind Turbines for On-Shore Sites. *Appl. Sci.* **2020**, *10*, 7595. [CrossRef]
7. Rehman, S.; Alam, M.; Alhems, L.M.; Rafique, M.M. Horizontal Axis Wind Turbine Blade Design Methodologies for Efficiency Enhancement—A Review. *Energies* **2018**, *11*, 506. [CrossRef]
8. Rehman, S.; Rafique, M.M.; Alam, M.M.; Alhems, L.M. Vertical axis wind turbine types, efficiencies, and structural stability—A Review. *Wind. Struct.* **2019**, *29*, 15–32.
9. Yurusen, N.Y.; Rowley, P.N.; Watson, S.J.; Melero, J.J. Automated wind turbine maintenance scheduling. *Reliab. Eng. Syst. Saf.* **2020**, *200*, 106965. [CrossRef]
10. Bakhshi, R.; Sandborn, P. Overview of Wind Turbine Field Failure Databases: A Discussion of the Requirements for an Analysis. In Proceedings of the ASME 2018 Power Conference Collocated with the ASME 2018 12th International Conference on Energy Sustainability and the ASME 2018 Nuclear Forum, Lake Buena Vista, FL, USA, 24–28 June 2018. [CrossRef]
11. Liu, W.; Zhang, W.; Han, J.; Wang, G. A new wind turbine fault diagnosis method based on the local mean decomposition. *Renew. Energy* **2012**, *48*, 411–415. [CrossRef]
12. Feng, Z.; Liang, M.; Zhang, Y.; Hou, S. Fault diagnosis for wind turbine planetary gearboxes via demodulation analysis based on ensemble empirical mode decomposition and energy separation. *Renew. Energy* **2012**, *47*, 112–126. [CrossRef]
13. Chen, J.; Pan, J.; Li, Z.; Zi, Y.; Chen, X. Generator bearing fault diagnosis for wind turbine via empirical wavelet transform using measured vibration signals. *Renew. Energy* **2016**, *89*, 80–92. [CrossRef]
14. Wenyi, L.; Zhenfeng, W.; Jiguang, H.; Guangfeng, W. Wind turbine fault diagnosis method based on diagonal spectrum and clustering binary tree SVM. *Renew. Energy* **2013**, *50*, 1–6. [CrossRef]
15. Tang, B.; Song, T.; Li, F.; Deng, L. Fault diagnosis for a wind turbine transmission system based on manifold learning and Shannon wavelet support vector machine. *Renew. Energy* **2014**, *62*, 1–9. [CrossRef]
16. Gao, Q.; Liu, W.; Tang, B.; Li, G. A novel wind turbine fault diagnosis method based on intergral extension load mean decomposition multiscale entropy and least squares support vector machine. *Renew. Energy* **2018**, *116*, 169–175. [CrossRef]
17. Lei, J.; Liu, C.; Jiang, D. Fault diagnosis of wind turbine based on Long Short-term memory networks. *Renew. Energy* **2019**, *133*, 422–432. [CrossRef]
18. Jiang, G.; He, H.; Yan, J.; Xie, P. Multiscale Convolutional Neural Networks for Fault Diagnosis of Wind Turbine Gearbox. *IEEE Trans. Ind. Electron.* **2019**, *66*, 3196–3207. [CrossRef]
19. Stetco, A.; Dinmohammadi, F.; Zhao, X.; Robu, V.; Flynn, D.; Barnes, M.; Keane, J.; Nenadic, G. Machine learning methods for wind turbine condition monitoring: A review. *Renew. Energy* **2019**, *133*, 620–635. [CrossRef]
20. Teng, W.; Ding, X.; Cheng, H.; Han, C.; Liu, Y.; Mu, H. Compound faults diagnosis and analysis for a wind turbine gearbox via a novel vibration model and empirical wavelet transform. *Renew. Energy* **2019**, *136*, 393–402. [CrossRef]
21. Cai, W.; Wang, Z. Application of an Improved Multipoint Optimal Minimum Entropy Deconvolution Adjusted for Gearbox Composite Fault Diagnosis. *Sensors* **2018**, *18*, 2861. [CrossRef]
22. Teng, W.; Ding, X.; Zhang, X.; Liu, Y.; Ma, Z. Multi-fault detection and failure analysis of wind turbine gearbox using complex wavelet transform. *Renew. Energy* **2016**, *93*, 591–598. [CrossRef]
23. Wang, X.; Tang, G.; He, Y. Compound fault diagnosis of wind turbine bearings based on COT-MCKD-STH under variable speed conditions. *J. Chin. Soc. Power Eng.* **2019**, *2019*, 220–226, (In Chinese with English abstract).
24. Wang, Y.; Tang, B.; Meng, L.; Hou, B. Adaptive Estimation of Instantaneous Angular Speed for Wind Turbine Planetary Gearbox Fault Detection. *IEEE Access* **2019**, *7*, 49974–49984. [CrossRef]
25. Wang, Z.; He, H.; Wang, J.; Du, W. Application Research of a Novel Enhanced SSD Method in Composite Fault Diagnosis of Wind Power Gearbox. *IEEE Access* **2019**, *7*, 154986–155001. [CrossRef]
26. Wang, Z.; Wang, J.; Cai, W.; Zhou, J.; Du, W.; Wang, J.; He, G.; He, H. Application of an Improved Ensemble Local Mean Decomposition Method for Gearbox Composite Fault Diagnosis. *Complexity* **2019**, *2019*, 1–17. [CrossRef]
27. Xiang, L.; Su, H.; Li, Y. Research on Extraction of Compound Fault Characteristics for Rolling Bearings in Wind Turbines. *Entropy* **2020**, *22*, 682. [CrossRef] [PubMed]

28. Zhong, J.-H.; Zhang, J.; Liang, J.; Wang, H. Multi-Fault Rapid Diagnosis for Wind Turbine Gearbox Using Sparse Bayesian Extreme Learning Machine. *IEEE Access* **2018**, *7*, 773–781. [CrossRef]
29. Gilles, J. Empirical Wavelet Transform. *IEEE Trans. Signal Process.* **2013**, *61*, 3999–4010. [CrossRef]
30. Gilles, J.; Heal, K. A parameterless scale-space approach to find meaningful modes in histograms—Application to image and spectrum segmentation. *Int. J. Wavelets Multiresolution Inf. Process.* **2014**, *12*. [CrossRef]
31. Géron, A. Support vector machines. In *Hands-On Machine Learning with Scikit-Learn and TensorFlow*; O'Reilly Media Inc.: Boston, MA, USA, 2017; pp. 145–165.
32. Keerthi, S.S.; Lin, C.-J. Asymptotic Behaviors of Support Vector Machines with Gaussian Kernel. *Neural Comput.* **2003**, *15*, 1667–1689. [CrossRef] [PubMed]
33. Apostolidis-Afentoulis, V.; Lioufi, K.I. SVM Classification with Linear and RBF Kernels. *Academia* **2015**. [CrossRef]
34. Mirjalili, S.; Mirjalili, S.M.; Lewis, A. Grey Wolf Optimizer. *Adv. Eng. Softw.* **2014**, *69*, 46–61. [CrossRef]
35. ElHariri, E.; El-Bendary, N.; Hassanien, A.E.; Abraham, A. Grey wolf optimization for one-against-one multi-class support vector machines. In Proceedings of the 2015 7th International Conference of Soft Computing and Pattern Recognition (SoCPaR2015), Fukuoka, Japan, 15 June 2016; pp. 7–12. [CrossRef]

Article

Subway Gearbox Fault Diagnosis Algorithm Based on Adaptive Spline Impact Suppression

Zhongshuo Hu [1,2], Jianwei Yang [1,2,*], Dechen Yao [1,2], Jinhai Wang [1,2] and Yongliang Bai [2,3]

[1] School of Mechanical-Electronic and Vehicle Engineering, Beijing University of Civil Engineering and Architecture, Beijing 100044, China; huzhongshuo@163.com (Z.H.); yaodechen@bucea.edu.cn (D.Y.); wangjinhai@bucea.edu.cn (J.W.)
[2] Beijing Key Laboratory of Performance Guarantee on Urban Rail Transit Vehicles, Beijing University of Civil Engineering and Architecture, Beijing 100044, China; yongliangbai@126.com
[3] School of Mechanical, Electronic and Control Engineering, Beijing Jiaotong University, Beijing 100044, China
* Correspondence: yangjianwei@bucea.edu.cn

Citation: Hu, Z.; Yang, J.; Yao, D.; Wang, J.; Bai, Y. Subway Gearbox Fault Diagnosis Algorithm Based on Adaptive Spline Impact Suppression. *Entropy* 2021, 23, 660. https://doi.org/10.3390/e23060660

Academic Editors: Yongbo Li, Fengshou Gu and Xihui (Larry) Liang

Received: 27 April 2021
Accepted: 19 May 2021
Published: 25 May 2021

Publisher's Note: MDPI stays neutral with regard to jurisdictional claims in published maps and institutional affiliations.

Copyright: © 2021 by the authors. Licensee MDPI, Basel, Switzerland. This article is an open access article distributed under the terms and conditions of the Creative Commons Attribution (CC BY) license (https://creativecommons.org/licenses/by/4.0/).

Abstract: In the signal processing of real subway vehicles, impacts between wheelsets and rail joint gaps have significant negative effects on the spectrum. This introduces great difficulties for the fault diagnosis of gearboxes. To solve this problem, this paper proposes an adaptive time-domain signal segmentation method that envelopes the original signal using a cubic spline interpolation. The peak values of the rail joint gap impacts are extracted to realize the adaptive segmentation of gearbox fault signals when the vehicle was moving at a uniform speed. A long-time and unsteady signal affected by wheel–rail impacts is segmented into multiple short-term, steady-state signals, which can suppress the high amplitude of the shock response signal. Finally, on this basis, multiple short-term sample signals are analyzed by time- and frequency-domain analyses and compared with the nonfaulty results. The results showed that the method can efficiently suppress the high-amplitude components of subway gearbox vibration signals and effectively extract the characteristics of weak faults due to uniform wear of the gearbox in the time and frequency domains. This provides reference value for the gearbox fault diagnosis in engineering practice.

Keywords: gearbox; signal interception; peak extraction; cubic spline interpolation envelope

1. Introduction

The gearbox is an important part of an urban rail train, and a malfunctioning gearbox will significantly affect the smoothness and comfort of the train operation. Furthermore, the service life of the whole structure will be reduced due to the long-term wear of gear damage [1]. At present, in actual train operation management and maintenance, most of the gearbox fault diagnosis is completed during the maintenance of the unit. To find a gearbox fault as early as possible, the real vehicle signals must be processed and diagnosed [2]. However, for the signal processing of a real vehicle gearbox, the working environment of the gearbox is complex and easily affected by vibrations of all kinds of connected parts. This can lead to various problems, such as spectrum distortion caused by the impact of the rail joint gap, the failure frequency not being obvious due to the uniform wear of the gearbox, and the speed of the train and frequency of gearbox being in an uncertain state. The traditional signal processing methods have difficulty capturing the fault characteristics of a gearbox. Therefore, filtering the interference in a real vehicle signal and accurately extracting the fault characteristic information from the vibration time-domain signal are the key problems to be solved in the fault diagnosis of subway gearboxes.

In recent years, scholars have proposed a series of methods for processing the fault signals of nonstationary stochastic processes, and many algorithms and processing ideas have been improved and applied to gearbox fault diagnosis. As a previously proposed signal decomposition method, empirical mode decomposition (EMD) [3], has been gradually developed in recent years, the optimized form of EMD is now relatively mature and widely used in fault diagnosis [4]. Ensemble empirical mode decomposition (EEMD) has been proposed to reduce the mode mixing phenomenon [5–10] in the process of EMD decomposition. This is realized by adding Gaussian white noise to change the characteristics of the extrema, and variants of this algorithm have been developed. Wu et al. [11] proposed a feature learning detection method based on EEMD and a Gaussian process classifier. Trelet was used for data dimension reduction as the Gaussian process input, and it is optimized by bacterial foraging optimization. In addition, complementary set empirical mode decomposition (CEEMD) introduces complementary noise [12–14], which eliminates redundant noise to a large extent in the reconstruction of signals, greatly shortening the processing time and improving the computational efficiency. Han et al. [15] used a combination of the Teager energy operator and CEEMD to extract features from the bearing fault signals of a wind turbine. This algorithm showed unique advantages in detecting the impact characteristics of signals and effectively extracted the fault features of low-speed bearings. Li et al. [16] used the improved complete CEEMD with adaptive noise method to decompose bearing vibration signals, extracted nonlinear entropy features, and built a multiclass intelligent recognition model based on an integrated support vector machine, effectively classifying experimental data under various operating conditions. Variational mode decomposition (VMD) is a kind of adaptive signal decomposition method. This method assumes that each mode is around a center frequency. The modal solution can be converted to a constrained optimization problem, drastically reducing the modal aliasing phenomenon [17]. Based on this method, many scholars modified the VMD algorithm for mechanical fault diagnosis [18–22]. Cai et al. [23] proposed a multipoint kurtosis–VMD compound algorithm that can reduce the original signal noise and extract recurrent failures by using multipoint kurtosis at the same time, and more cycle vectors were constructed to determine the decomposition layers K. Compared with the traditional particle swarm optimization algorithm and ant colony algorithm, this method shows a speed advantage and eventually effectively extracts the compound fault characteristics. Hua et al. [24] proposed an inherent mode function selection method based on the resonance frequency for VMD parameter optimization and selected the modal components with fault information according to the resonance frequency. The results showed that this method can extract weak signals of early rolling bearing faults and realize correct judgment of bearing faults. Liu et al. [25] improved the VMD algorithm based on an autoregression (AR) model. By reducing the interference of low-frequency components and denoising each component, Hilbert envelope analysis was used to demodulate the signal, and the fault was determined by combining the demodulation frequency and theoretical fault frequency. Miao et al. [26] proposed an improved parameter-adaptive VMD method to construct a new comprehensive kurtosis index, which takes into account periodicity and impacts. This algorithm is superior to the traditional VMD method and further extends the application of VMD for compound fault diagnosis. Many scholars have also applied neural networks for fault diagnosis [27–35], and the improved algorithms have shown good results. In addition, there are also many decomposition methods based on signal processing, including local mean decomposition [36–40], Fourier decomposition [41], and other methods, which are widely used in gearbox fault diagnosis. Lei et al. [42] combined the advantages of integrated local mean decomposition and fast spectral kurtosis to carry out fault detection of rotating machinery and finally verified the effectiveness of this algorithm for the fault diagnosis of gearboxes and rolling bearings. Dou et al. [43] proposed a mechanical fault feature extraction method based on Fourier decomposition, which has the characteristics of adaptive narrowband filtering at high and low frequencies. In terms of separating low-frequency signal components, due to the traditional EMD method, it has a good effect when used for the feature extraction of mechanical vibration signals. Pang et al. [44] proposed an enhanced singular spectrum decomposition (ESSD) method that highlights fault signal components by introducing differentiation and integration operators. This method exhibits strong anti-interference abilities and better performances when processing experimental signals compared with traditional singular spectrum decomposition (SSD) and VMD.

For the research on wheel–rail impact caused by rail joints, Yang et al. [45] designed an experiment and finite element model to reproduce the noise and vibration signals of a wheelset passing the rail joints. The results show that the vibration energy of the vertical impact is mainly concentrated around 300 and 1000 Hz, while the dominant frequency range of transverse vibration is between 300 and 1200 Hz. Tajalli et al. [46] used an experiment and finite element model to study different relative positions of sleeper and rail joint influence on wheel–rail impact; the results show that the wheel–rail contact force is 1.77 times the size of the static load when wheelset passes the rail joint, which caused the wheel–rail vibration acceleration signal to change sharply. The research also points out that the impact force can be reduced by 40% when using a fishplate to connect the rail joint. Choi et al. [47] used the ANSYS model to analyze the wheel–rail impact force at the rail joint, and the model was verified by experimental data. They evaluated the wheel–rail contact force by comparing track impact factor (TIF), and the results showed that TIF value with rail joints is 57% higher than that for continuous welded rails. This research shows that rail joints produce wheel–rail impact, which will bring interference components to gearbox diagnosis. Real vehicle signal data are very scarce currently, and this problem is proposed in real vehicle signal processing. The traditional EMD and VMD methods have remarkable performance on the bearing vibration signals collected in the laboratory, but the effect on the wheel–rail impact filtering in the measured train gearbox signals is not ideal; most of these algorithms were proposed based on laboratory data, and their feasibility in practical engineering needs to be further verified. The method proposed in this paper aims to eliminate the interference of wheel–rail impact on the spectrum more effectively.

To solve this problem, this paper proposes a time-domain adaptive interception algorithm based on a cubic spline interpolation envelope. The purpose is to remove wheel–rail impact interference in real train signals. The basic idea is to automatically extract the peak value from the whole large sample signal and calculate the train speed data corresponding to the short-term sample by using the peak interval. Based on the position of the peak value, the single large sample signal is adaptively intercepted and divided into several short-term samples, after which it is analyzed in the time–frequency domain and finally compared with healthy signals. The results show that this method can effectively eliminate the low-frequency interference caused by rail impacts, and it can calculate and complement the missing train speed data. Moreover, the processed signals can reflect certain fault rules in both the time-domain statistics and frequency-domain analysis, and the method can diagnose uniform wear of subway gearboxes.

The main contributions of this work are summarized as follows: A gearbox fault diagnosis algorithm based on adaptive spline is proposed for suppressing impact components in real vehicle signals adaptively. Speed data can also be calculated based on the proposed algorithm. A new feature representation method is used in the analysis part; it is based on multisample preponderance after processing by the proposed algorithm.

This work has the following organization: The foundational spline method and the proposed algorithm are both introduced in Section 2. Section 3 describes the acquisition of real vehicle signal and presents the results of the analysis of the proposed method carried out using real vehicle signal, compared with EEMD and VMD. Conclusions are drawn in Section 4.

2. Signal Segmentation Method Based on Cubic Spline Interpolation Envelope

The impact components will seriously interfere with the low frequency after Fourier transform, which causes severe distortion of the spectrum. At present, there may be some difficulties in finding the peak and intercept signal adaptively. In this section, a time-domain adaptive interception algorithm based on cubic spline interpolation envelope is proposed to deal with interference of wheel–rail impact in real vehicle signals. A long-time large sample signal is divided into several short-term small sample signals, and impact components are extracted and eliminated at the same time.

2.1. Cubic Spline Interpolation Envelope

The cubic spline interpolation method has good convergence properties. In the process of adaptive signal segmentation, it exhibits good stability. Cubic spline interpolation uses multiple piecewise cubic polynomials and is piecewise continuous, so its first and second derivative functions are continuous, and there are no errors at the nodes or at the nodes of the first and second derivatives.

Cubic spline interpolation is defined as a partition $x_0 < x_1 < \ldots < x_{n-1} < x_n$ given in the interval $[x_0, x_n]$. The function $L(x)$ is defined on each segment interval $[x_{i-1}, x_i](i = 1, 2, \ldots, n)$. $L(x)$ is a cubic polynomial, and in the entire interval $[x_0, x_n]$, $L(x)$ is a second-order continuous and differentiable function. Each polynomial satisfies the following at each node $x_i (i = 1, 2, \ldots, n-1)$:

$$L^{(k)}(x_i - 0) = L^{(k)}(x_i + 0), k = 0, 1, 2, \quad (1)$$

The constraint conditions are as follows:

$$\begin{cases} L(x_i) = y_i, (i = 0, 1, \ldots, n) \\ L(x_i - 0) = L(x_i + 0), (i = 1, 2, \ldots, n-1) \\ L'(x_i - 0) = L'(x_i + 0), (i = 1, 2, \ldots, n-1) \\ L''(x_i - 0) = L''(x_i + 0), (i = 1, 2, \ldots, n-1) \end{cases}, \quad (2)$$

In the cubic spline interpolation method, there are three types of boundary conditions: clamped, natural, and not-a-knot. Since the second derivative of cubic polynomial $L(x)$ is a first-order function, let $[L''(x_0), L''(x_1), \ldots, L''(x_n)] = [Q_0, Q_1, \ldots, Q_n]$, and then the interpolation function of each piecewise interval is as follows:

$$L(x) = \frac{Q_{i-1}}{6h_i}(x_i - x)^3 + \frac{Q_i}{6h_i}(x - x_{i-1})^3 + (\frac{y_i - y_{i-1}}{h_i} - \frac{Q_i - Q_{i-1}}{6}h_i)x \\ + y_i - \frac{Q_i}{6}h_i^2 - (\frac{y_i - y_{i-1}}{h_i} - \frac{Q_i - Q_{i-1}}{6}h_i)x_i \quad (3)$$

In Equation (3), $h_i = x_i - x_{i-1}$.

With the support of the above principle, the original signal is processed by a cubic spline interpolation envelope. In the whole signal, the amplitude sequence is x, the corresponding time series is t, and $n+1$ signal sequence points are denoted as (t_0, x_0), $(t_1, x_1) \ldots (t_i, x_i) \ldots (t_n, x_n)$. The step size to find the maximum number of points n_p is defined, and the maximum value from the fixed signal step size is output. The signal length is optimized based on experience. An excessive step size will cause significant distortion, while too small of a step size will increase the number of calculations. Therefore, it is necessary to optimize the selection of step size, as shown in Figure 1, which is the amount of peak value under different step sizes. It can be seen that when the step size reaches a certain range, the amount of peak value tends to be stable several times. In the first stabilization, the average of corresponding values will be the best n_p (red mark), and then the coordinate of each maximum point is $(a_j, b_j), j = 0, \ldots, n$.

Using the maximum output point (a_j, b_j) in the above fixed step size n_p as the interpolation point, Equation (3) shows that the cubic spline interpolation function on the interval $[a_j, a_{j+1}]$ is expressed as follows:

$$L(t) = \frac{Q_{j-1}}{6h_j}(a_j - t)^3 + \frac{Q_j}{6h_j}(t - a_{j-1})^3 + (\frac{b_j - b_{j-1}}{h_j} - \frac{Q_j - Q_{j-1}}{6}h_j)t \\ + b_j - \frac{Q_j}{6}h_j^2 - (\frac{b_j - b_{j-1}}{h_j} - \frac{Q_j - Q_{j-1}}{6}h_j)a_j \quad (4)$$

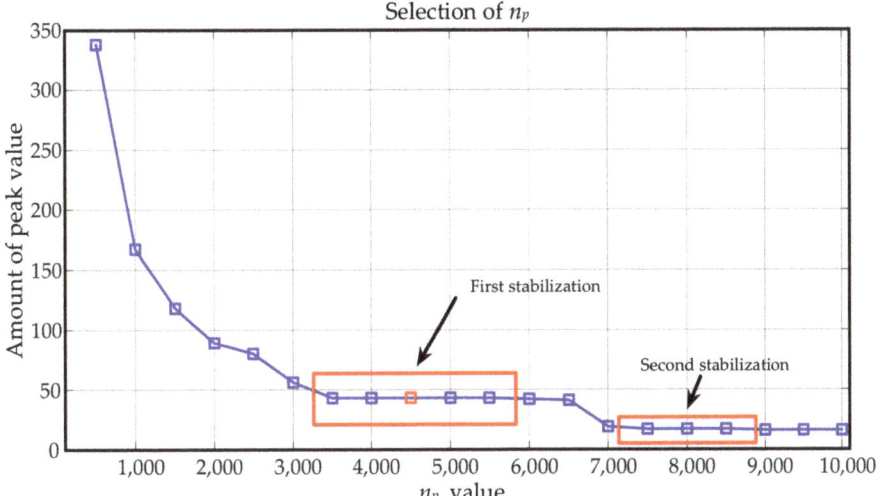

Figure 1. The selection of n_p value.

In Equation (4), $h_j = a_j - a_{j-1}$. A fixed boundary condition is selected; that is, the first derivative value at each maximum coordinate point a_0 and a_j is $L'(a_0) = b'_0 = L'(a_n) = b'_n = 0$, and the solution of Q_j is as follows:

$$\begin{bmatrix} 2 & 1 & & & \\ \mu_2 & 2 & 1-\mu_2 & & \\ & \ddots & \ddots & \ddots & \\ & & \mu_{n-1} & 2 & 1-\mu_{n-1} \\ & & & 1 & 2 \end{bmatrix} \begin{bmatrix} Q_1 \\ Q_2 \\ \vdots \\ Q_{n-1} \\ Q_n \end{bmatrix} = \begin{bmatrix} \lambda_1 \\ \lambda_2 \\ \vdots \\ \lambda_{n-1} \\ \lambda_n \end{bmatrix}, \quad (5)$$

where $\mu_j = \frac{h_j}{h_j+h_{j+1}}$, $\lambda_1 = \frac{6}{h_1}(\frac{b_1-b_0}{h_1} - b'_0)$, $\lambda_n = \frac{6}{h_n}(b'_n - \frac{b_n-b_{n-1}}{h_n})$, and $\lambda_j = 6(\frac{b_{j+1}-b_j}{h_{j+1}} - \frac{b_j-b_{j-1}}{h_j})\frac{1}{h_j+h_{j+1}}$ $(j = 2, 3, \ldots, n-1)$.

Through the catch-up method, $[Q_0, Q_1, \ldots, Q_n]$ can be solved for using Equation (5), which can be back-substituted into Equation (5) to obtain the cubic spline interpolation envelope of the original signal. At this time, the envelope curve has filtered out a significant amount of peak interference other than the wheel–rail impact, and the maximum value of the envelope curve can be determined and output as (G_i, H_i), $i = 0, 1, \ldots, n$.

2.2. Impact Component Extraction and Short-Term Signal Sample Segmentation

There is some error between the crude extracted coordinates and the original signal due to the envelope curve fitting. Therefore, to eliminate this error, the original signal peak is extracted. Because the time interval between the wheel and rail impact has a certain regularity, that is, the ratio of the rail length to vehicle speed, the abscess interval of the pulse can be estimated as $\Delta t \approx 1.4$ s. The search interval $[G_i - \frac{\Delta t}{2} \bullet f_s, G_i + \frac{\Delta t}{2} \bullet f_s]$ is set according to the abscissa of the maximum value of the envelope curve G_i, and the maximum values $(G_i', (H_i', i = 0, 1, \ldots, n$ are found within the search interval to complete the extraction steps of the whole signal.

After the impact component is extracted, the x-coordinate difference of maximum values (G_i', H_i'), which is $\Delta G = G_i' - G_{i-1}'$, is calculated; ΔG is the interval time of the train passing along a steel rail. To control the length of each short-term sample and reduce the number of variables, a 1 s signal is intercepted between the extracted wheel–rail impact interval to complete the overall signal segmentation.

3. Experiment and Results

3.1. Real Vehicle Data Collection

The vibration acceleration data of a real vehicle gearbox were collected. Test information is as follows:

Data acquisition equipment: three-axis vibration sensor (measurement range is 500 g), MDR-80 dynamic acquisition instrument, MDR-80 mobile data recording system.

Test object: subway vehicle.

Test location: a whole Beijing subway line, total distance of single acquisition is about 81 km.

Figure 2a shows the field sensor layout, and the vibration sensor was installed on the outer shell of the gearbox. The sensor adopted a triaxial vibration sensor, and the sampling frequency was set to 10 kHz. Figure 2b shows the data acquisition equipment. The acceleration and deceleration processes of the vehicle were kept as uniform as possible. When driving at a constant speed on a straight line segment, the speed should be kept at about 67 km/h, and the stopping time of each stop should be 2 s. The data of the faulty and healthy gearbox were collected on the same subway line twice. Two data samples were collected, and the single sample collection time lasted about 120 min. The total number of sample points was above 70 million for one sample. The tooth ratio of the measured gearbox was 100:13. The surface damage of the internal gear of the fault gearbox is shown in Figure 2c, and there were cracks, peeling, and wear on the gear surface. The other calculation parameters are shown in Table 1.

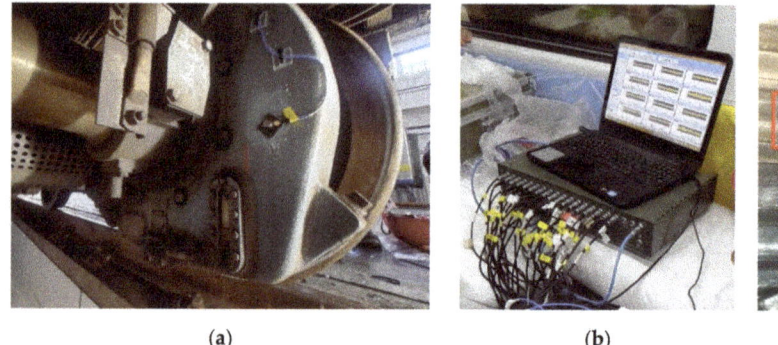

(a) (b) (c)

Figure 2. Data acquisition of real vehicle gearbox: (**a**) gearbox sensor layout; (**b**) data acquisition equipment; (**c**) tooth surface fault.

Table 1. Train speed calculation parameters.

	Length of Rail/l_r(m)	Wheel Diameter/d (mm)	Gear Ratio/i
Parameter	25	821	100:13

3.2. Results

Figure 3 shows a segment of the signal when the vehicle was operating at a uniform speed in the time domain. Wheel–rail impacts were distributed uniformly over the whole segment of the time-domain signal. After the Fourier transform, the spectral components were mixed, and the fault components could not be determined and separated. The fault frequency could not be distinguished, and there was more interference at low frequencies, which was caused to some extent by the frequency of rail joint gap impacts.

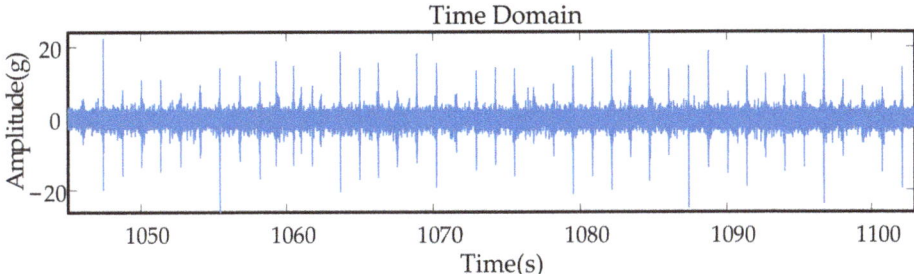

Figure 3. Time-domain diagram of original signals.

Tooth surface wear is an inevitable phenomenon over the life of a gearbox. The failure frequency was not fixed, the train speed varied during the signal measurement, and the gear switching frequency floated up and down. Figure 4 shows the healthy and faulty original signal spectrum graph. The original noise signal interference was evident, and various frequency components were mutually coupled. The signal characteristics of some weak faults were easily submerged by noise, so it was difficult to judge the fault situation based on the spectrum peaks.

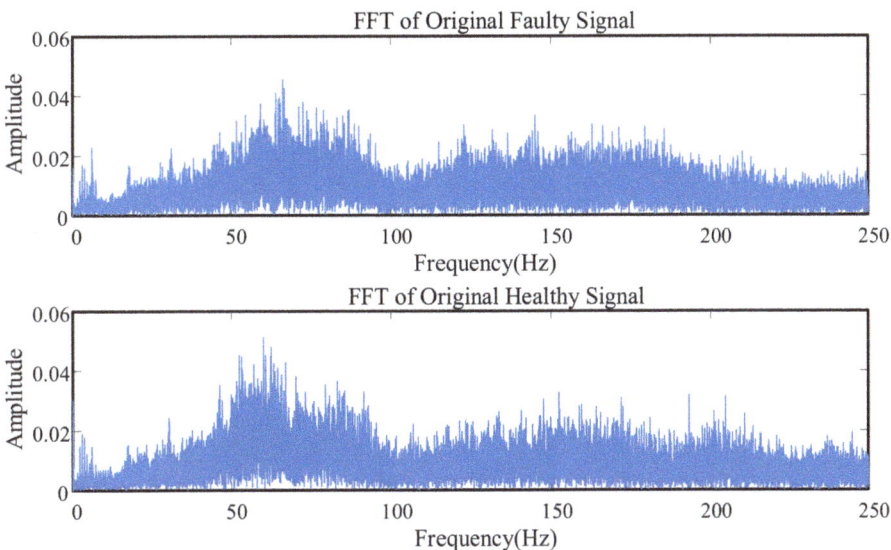

Figure 4. Spectrum comparison of healthy and faulty states.

Figure 5 shows the original signal power spectrum diagram, and the inset shows the meshing frequencies. The red line represents the faulty signal, and the blue line represents the healthy signal. The healthy and faulty states of the power spectrum in the frequency domain were similar, and the power of the healthy signal was even greater than that of the faulty signal in one region. A pattern of the differences between the two spectra is not evident, and the power spectrum analysis of the original signal cannot show the fault characteristics.

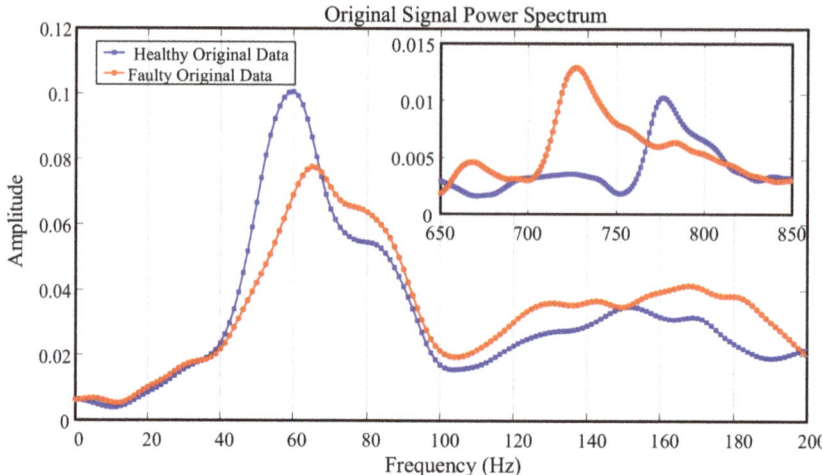

Figure 5. Power spectrum of original signal.

Based on the principle described above, the cubic spline interpolation envelope signal intercept method was applied. The real vehicle operating conditions and changes interfered with the signal, and the gearbox signal was mixed with the wheel–rail impact signal. Thus, direct calculation and analysis of the original signal could not provide better analysis of the composition of the gearbox fault.

The specific steps of the algorithm are as follows, and the algorithm flow chart is shown in Figure 6:

(1) The extremum step n_p is defined to find and output the maximum value (a_j, b_j) from the extremum step in the original signal.

(2) Using the cubic spline interpolation method, the envelope curve $S(t)$ is calculated for the above maximum array, and the maximum value (G_i, H_i) of the envelope is output.

(3) The maximum value (G_i', H_i') of the original signal is found within half of the abscess interval of the maximum value, which is the peak value of the original signal rail joint gap impact.

(4) According to the coordinates of the adjacent impact peak, some of the short-term sample signals between the peak values were intercepted, and a large number of short-term sample signals were intercepted for statistical analysis in the time–frequency domain.

In this way, the wheel–rail impact components in the original signal can be effectively removed, and the sample size of the data can also be increased, laying a foundation for the next step of the time-domain statistics. Figure 7 shows the whole signal segment. The signal after adaptive interception is shown, where the gray components are the original signal and the blue components are parts of the intercepted signal. The rail joint gap impact component was effectively eliminated, and a gearbox signal component with a more uniform amplitude was retained. Wheel–rail impact components are marked in red circles; their accuracy directly determines the effectiveness of the algorithm. After many instances of verification and calculation, the accuracy rate was about 98%, and errors often occurred at the boundary part of signals.

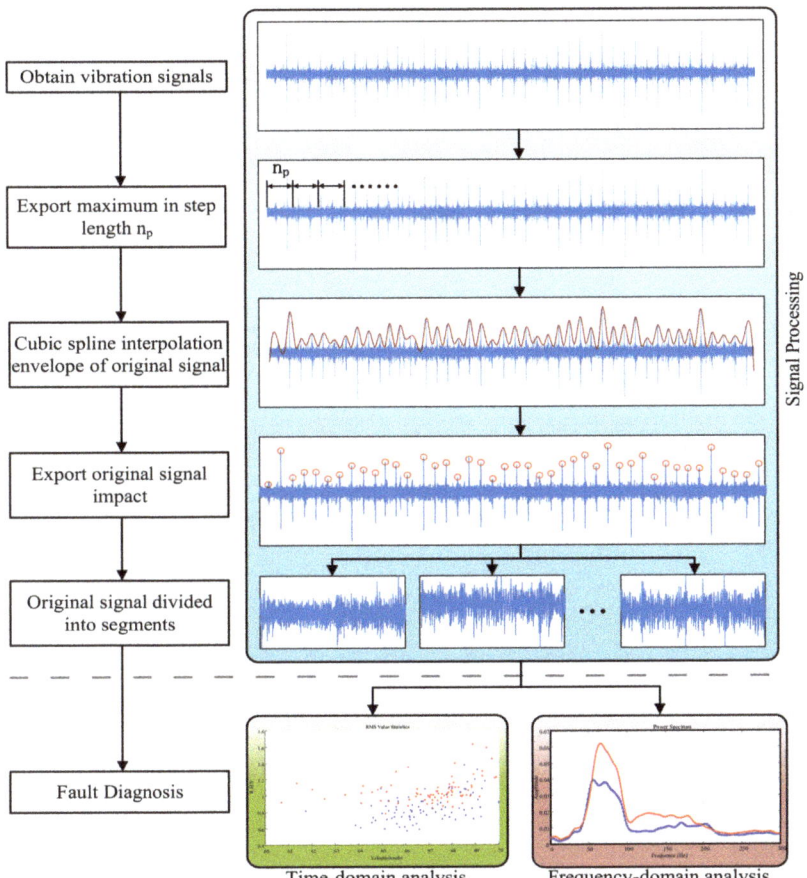

Figure 6. Algorithm flow chart.

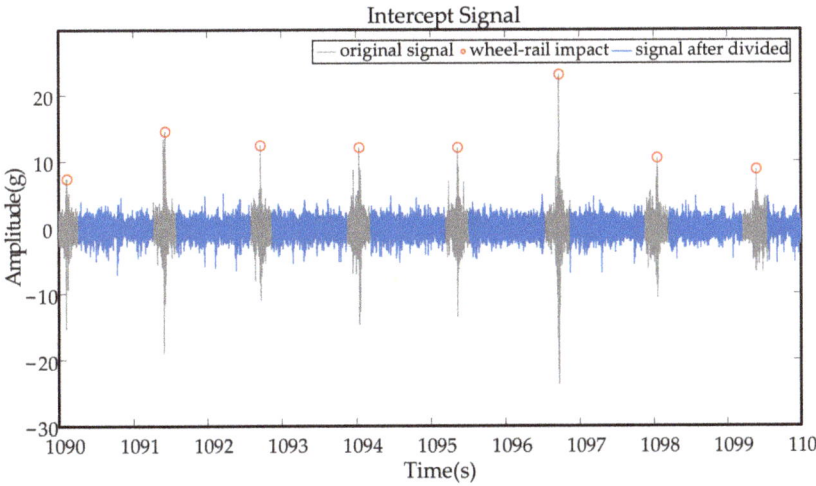

Figure 7. Schematic diagram of signal interception.

Since the original measured signals often lack the speed information of the train when it is operating, the calculation of the speed data is completed according to the extracted time between the wheel–rail impact peak and the length of the rail. The parameters such as the length of the rail and the measured wheel diameter are shown in Table 1, and the train speed is $v = \frac{l_r}{\Delta G}$. It is also possible to calculate the pinion frequency $f_r = \frac{v}{\pi d}$, the large gear frequency $f_R = f_r \bullet i$, and the meshing frequency $f_m = 100 f_r = 13 f_R$. It should be noted that the speed calculated at this time is the average speed of the train passing a whole section of rail, as shown in Figure 8. After the signal interception, there are some incorrect calculated speed data. Due to changes in speed and rail length, the distance between impacts may become irregular, which will make calculated speed data invalid. To ensure that the data were collected when the train moved at as constant a speed as possible, the calculated speed was used as the discriminant standard. Outlying speed points were screened out by setting a threshold: the maximum acceleration could not be greater than 1.5 m/s^2 when the train was running. In the figure, the slope of the curve pointed to by the arrow is too steep, so points on either side of this point will be regarded as outlying speed data, and short-term samples on either side of the point will not be analyzed in the time–frequency domain. At this point, the entire adaptive signal segmentation processing was completed.

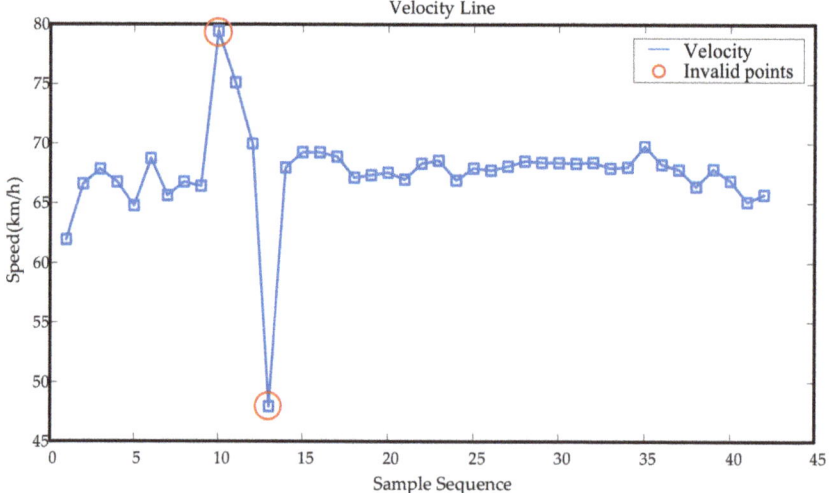

Figure 8. Velocity line diagram.

3.3. Statistical Analysis of Time-Domain Root Mean Square Values

First, the state of the gearbox was judged in the time-domain analysis. Since a single large sample was divided into several short-term samples after signal segmentation, the time-domain statistical analysis was carried out with the advantage of multiple samples, and the main calculated parameter was the root mean square (RMS) value. The time-domain index RMS of each signal after segmentation was calculated. The RMS is plotted versus the train speed for both the healthy and faulty states in Figure 9. The blue points are the healthy state data, the red points are the faulty state data, the red line is the faulty data envelope, the blue line is the healthy data envelope, and the dark red and dark blue X marks are the faulty and healthy data scatter centers of mass, respectively. From the perspective of the root mean square value, the overall level and average level of the faulty state were 30% higher than those of the healthy state. Combined with the speed distribution, although the root mean square value from low speed to high speed had a slight linear increase, the faulty state level was always above the healthy state.

The Jaccard similarity coefficient of the two types of cluster areas is used to indicate the similarity degree of the clusters. The formula of the Jaccard coefficient is as follows:

$J(S_H, S_F) = \frac{|S_H \cap S_F|}{|S_H \cup S_F|}$, where S_H is the area of the convex envelope of the healthy scattered points and S_F is the area of the convex envelope of the faulty scattered points. As shown in Figure 10, the Jaccard similarity coefficient of the two clusters was 33.87%, indicating that the similarity degree of the two types of data was extremely low. This further indicated that there was a significant difference between the healthy and faulty data, and the fault features of the time-domain root mean square statistics were prominent.

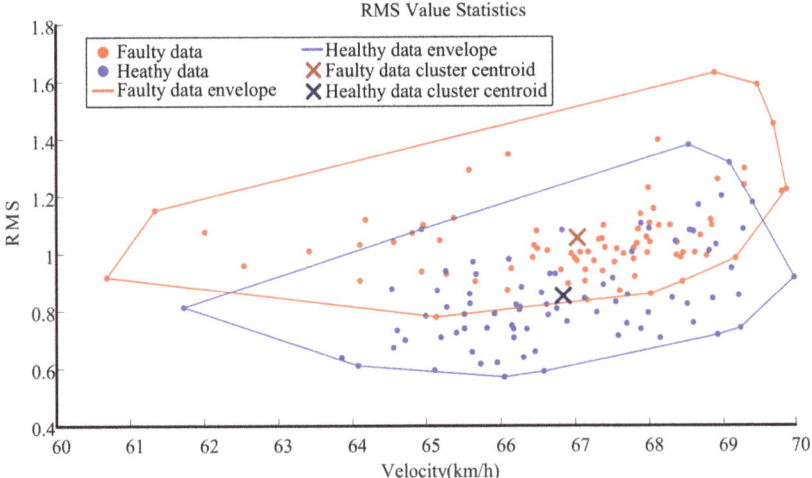

Figure 9. Statistics on root mean square values of short-term samples.

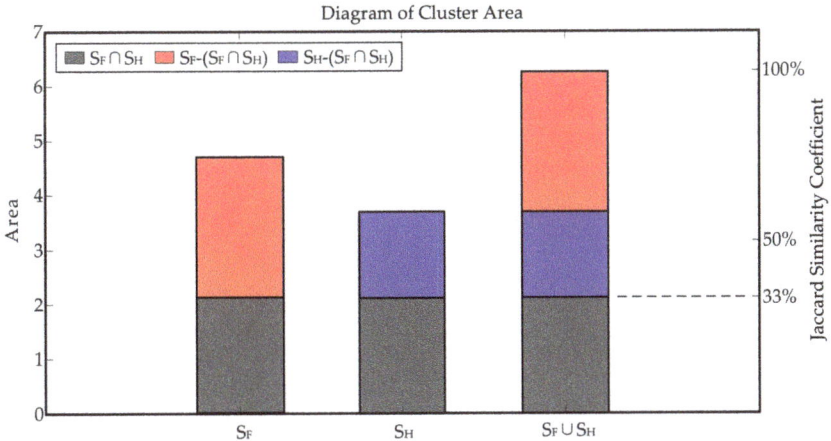

Figure 10. Cluster area and Jaccard similarity coefficient.

3.4. Multiple Mean Power Spectrum Analysis in Frequency Domain

After the fault differences were found in the time domain, the fault characteristics were further characterized in the frequency domain. According to the above analysis, since the train speed varied continuously, the average speed of the short-term sample train was 66 km/h. To determine the rotational frequency range of the large and small gears, the rotational frequencies of the small and large gears were 55.5 and 7.22 Hz, which were calculated using the wheel diameter and tooth ratio.

The signal segmentation algorithm proposed in this paper was used to divide the large sample signal into several short-term sample signals, and then the power spectrum of each small sample signal was calculated. To control the variables as much as possible, 35 short-term healthy samples and 35 short-term faulty samples were selected, and all the samples were collected from the same section on the rail line. The signal power spectrum is shown in Figure 11, where each curve represents the signal power spectral density for a short-term sample. Figure 11a,b show the results for healthy and faulty gearboxes, respectively. Most of the peak values of the faulty sample were concentrated at small gear rotation frequencies. Relative to the healthy data, the curve of the faulty data was steeper, and the peak frequency range was more concentrated. However, the above characteristics were not prominent in the large gear rotational frequency range.

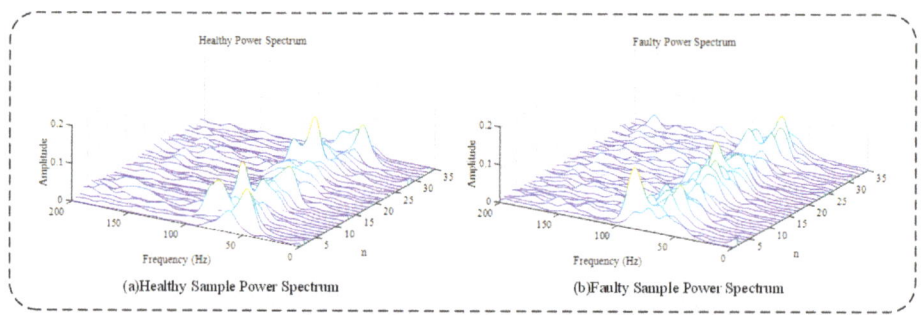

Figure 11. Power spectrum of each short-term sample.

To represent the differences in the characteristics, the average power spectra of the faulty and healthy states were obtained, as shown in Figure 12. The main plot shows the power spectrum of the gear rotational frequency, and the inset shows the power spectrum of the meshing frequency. The blue line represents the healthy state, and the red line represents the faulty state. The power spectral density function of the fault near the gear rotational frequency was much higher than that of the healthy state, especially in the region corresponding to the double small gear rotational frequency. Due to cracks and spalling failures in the gear surface and the changing speed of the train, the frequency-domain features had no specific fault frequency, which confirmed the fault features of multiple cracks and wear on the gearbox tooth surface. Thus, the fault diagnosis of the measured gearbox data was complete.

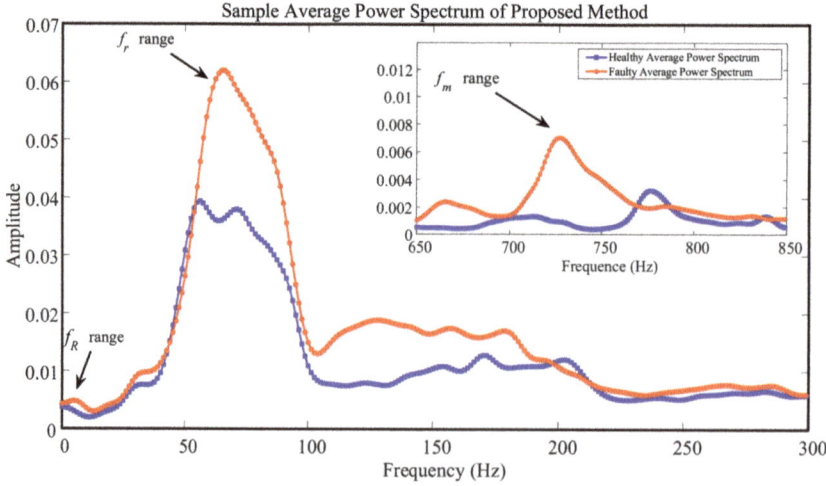

Figure 12. Average power spectrum of short-term samples with proposed method.

EEMD and VMD were selected to be compared with the proposed method, and the signal was divided into several 10 s short-term signals. After decomposed by using the two methods, IMFs were obtained, as shown in Figure 13. The IMFs with larger kurtosis were selected for reconstruction. The power spectra of reconstructed signals are shown in Figure 14. It can be clearly seen that EEMD and VMD failed to filter the impact components, which still affected the spectra significantly.

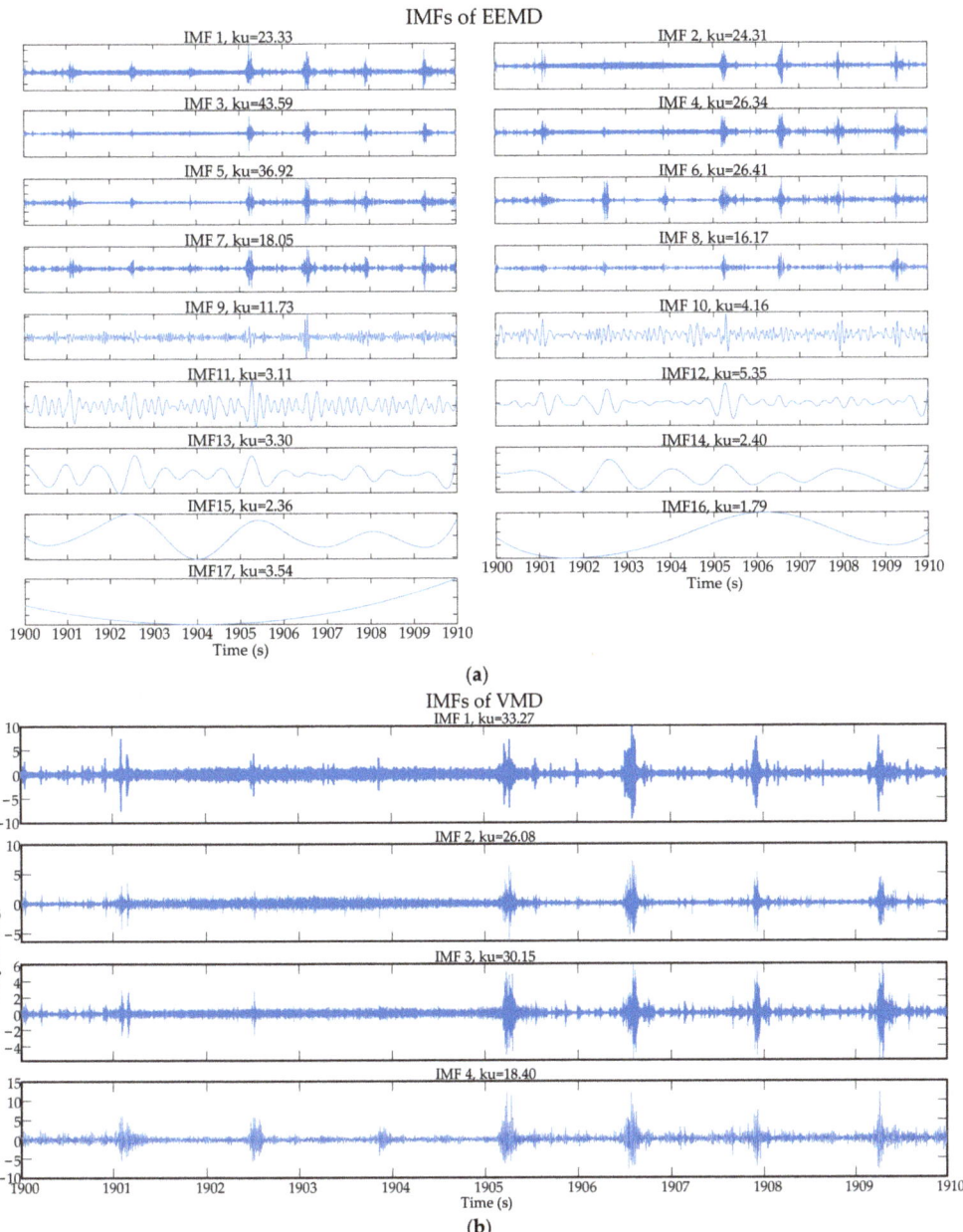

Figure 13. IMFs after decomposition by EEMD and VMD: (**a**) IMFs of EEMD; (**b**) IMFs of VMD.

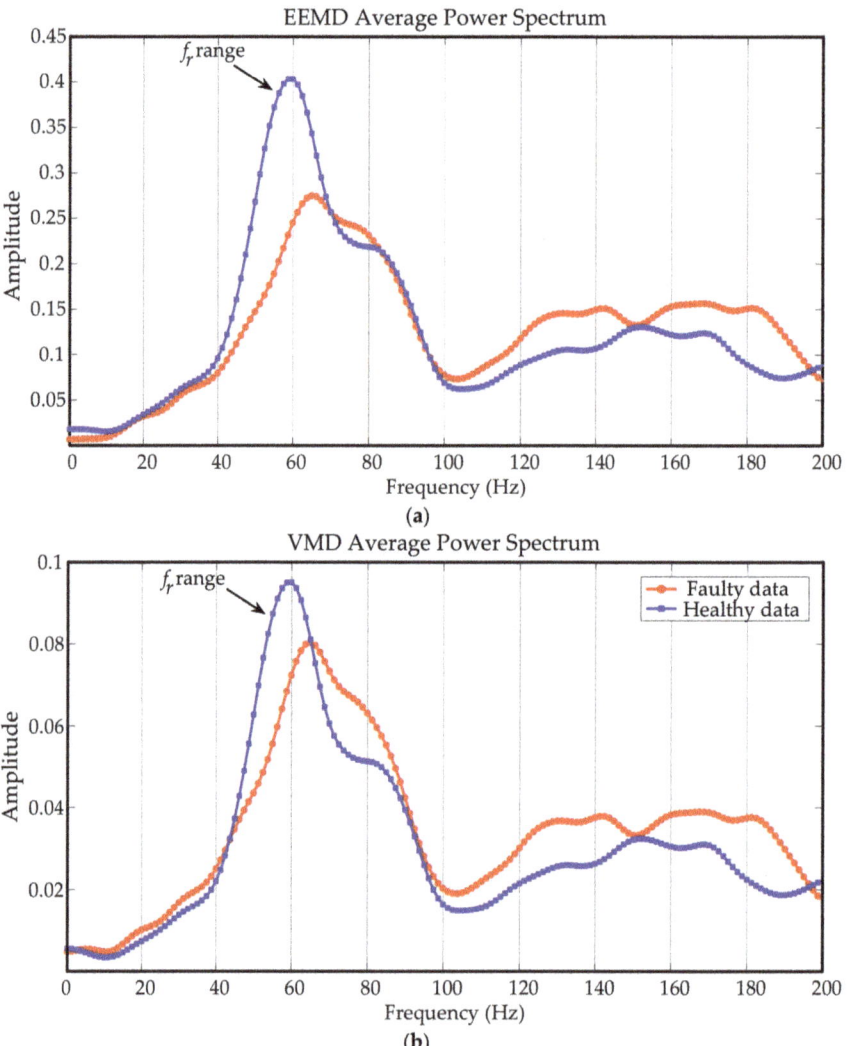

Figure 14. Average power spectra of compared methods: (a) EEMD average power spectrum; (b) VMD average power spectrum.

The difference between healthy and faulty power spectra was calculated to evaluate the methods' effects. As shown in Figure 15, a larger value at the f_r range means the method has a better effect. It can be seen that the value of EEMD is far less than 0, and VMD is slightly better than EEMD, but it is still in the state of less than 0, which means it failed to distinguish between faulty and healthy signals. The method proposed in this paper has the best effect, with a value that is much greater than 0. In summary, removing the wheel–rail impact interference from the original signal can effectively highlight the gearbox fault features with cracks, wear, and other damage.

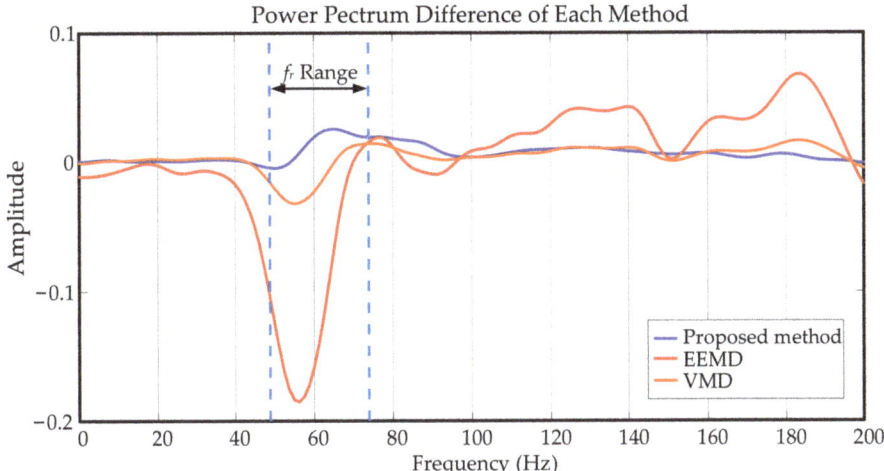

Figure 15. Power spectrum difference of each method.

4. Conclusions

In this paper, measured signals of real trains were analyzed. This work provides reference value for the fault diagnosis of train gearboxes in engineering practice. A signal segmentation algorithm was proposed, and the advantages of this algorithm are as follows:

(1) The impacts between the wheelsets and rail joint gaps in a real vehicle signal cause significant spectrum distortion in the low-frequency region. However, the signal also exhibits a certain regularity in the time domain. By using this regularity, this paper proposes a segmentation algorithm based on cubic spline interpolation. It can suppress the high amplitude of the shock response signal and divide a single large sample signal into several short-term sample signals.

(2) In the case of an uncertain train speed, this method can calculate the data for the train speed by using the extracted wheel–rail impacts, which provides a certain data basis for the calculation of the rotational frequency and fault frequency band in the frequency-domain analysis.

(3) Taking advantage of the number of samples after signal segmentation, statistical analysis of the short-term sample signal was completed in the time and frequency domains. The results showed that the algorithm has a strong ability to filter the disturbances caused by the impact of the rail joint gap. It can extract the useful short-term signal sample accurately. Compared to EEMD and VMD, the proposed algorithm can suppress and remove the impact components of the original real vehicle signal effectively. The results can detect regularities in the faulty signal and highlight the gearbox fault characteristics due to cracks, wear, and other damage in the time- and frequency-domain analyses.

Author Contributions: All authors conceived and designed the study. Software and paper writing, Z.H.; methodology, J.Y.; supervision, D.Y.; visualization, J.W.; data curation, Y.B. All authors have read and agreed to the published version of the manuscript.

Funding: This research was funded by the key project of 2020 Science and Technology Plan of Beijing Municipal Commission of Education under Grant KZ202010016025, and the National Natural Science Foundation of China under Grant 51975038.

Data Availability Statement: Data sharing not applicable.

Acknowledgments: The authors appreciate the support from the Beijing Key Laboratory of Performance Guarantee on Urban Rail Transit Vehicle for this research.

Conflicts of Interest: The authors declare no conflict of interest.

References

1. Yang, J.; Bai, Y.; Wang, J.; Zhao, Y. Tri-axial vibration information fusion model and its application to gear fault diagnosis in variable working conditions. *Meas. Sci. Technol.* **2019**, *30*, 095009. [CrossRef]
2. Chen, H.; Jiang, B.; Ding, S.; Huang, B. Data-driven fault diagnosis for traction systems in high-speed trains: A survey, challenges, and perspectives. *IEEE Trans. Intell. Transp. Syst.* **2020**. [CrossRef]
3. Huang, N.E.; Shen, Z.; Long, S.R.; Wu, M.C.; Shih, H.H.; Zheng, Q. The empirical mode decomposition and the hilbert spectrum for nonlinear and non-stationary time series analysis. *Proc. Math. Phys. Eng. Sci.* **1998**, *454*, 903–995. [CrossRef]
4. Zhang, W.; Zhou, J. A comprehensive fault diagnosis method for rolling bearings based on refined composite multiscale dispersion entropy and fast ensemble empirical mode decomposition. *Entropy* **2019**, *21*, 680. [CrossRef]
5. Wu, Z.; Huang, N.E. Ensemble empirical mode decomposition: A noise-assisted data analysis method. *Adv. Adapt. Data Anal.* **2011**, *1*, 1–41. [CrossRef]
6. Sadegh, H.M.; Esmaeilzadeh, K.S.; Saleh, S.M. Quantitative diagnosis for bearing faults by improving ensemble empirical mode decomposition. *ISA Trans.* **2018**, *83*, 261–275.
7. Hou, J.; Wu, Y.; Gong, H.; Ahmad, A.S.; Liu, L. A novel intelligent method for bearing fault diagnosis based on eemd permutation entropy and gg clustering. *Appl. Sci.* **2020**, *83*, 386. [CrossRef]
8. Ge, J.; Niu, T.; Xu, D.; Yin, G.; Wang, Y. A rolling bearing fault diagnosis method based on eemd-wsst signal reconstruction and multi-scale entropy. *Entropy* **2020**, *22*, 290. [CrossRef] [PubMed]
9. Prosvirin, A.E.; Islam, M.; Kim, J.M. An improved algorithm for selecting imf components in ensemble empirical mode decomposition for domain of rub-impact fault diagnosis. *IEEE Access* **2019**, *7*, 121728–121741. [CrossRef]
10. Nguyen, V.H.; Cheng, J.S.; Yu, Y.; Thai, V.T. An architecture of deep learning network based on ensemble empirical mode decomposition in precise identification of bearing vibration signal. *J. Mech. Sci. Technol.* **2019**, *33*, 41–50. [CrossRef]
11. Wu, E.Q.; Wang, J.; Peng, X.Y.; Zhang, P.; Law, R.; Chen, X. Fault diagnosis of rotating machinery using gaussian process and eemd-treelet. *Int. J. Adapt. Control Signal Process.* **2019**, *33*, 52–73. [CrossRef]
12. Yeh, J.R.; Shieh, J.S.; Huang, N.E. Complementary ensemble empirical mode decomposition: A novel noise enhanced data analysis method. *Adv. Adapt. Data Anal.* **2010**, *2*, 135–156. [CrossRef]
13. Wang, L.; Shao, Y. Fault feature extraction of rotating machinery using a reweighted complete ensemble empirical mode decomposition with adaptive noise and demodulation analysis. *Mech. Syst. Signal Process.* **2020**, *138*, 106545.1–106545.20. [CrossRef]
14. Kou, Z.; Yang, F.; Wu, J.; Li, T. Application of iceemdan energy entropy and afsa-svm for fault diagnosis of hoist sheave bearing. *Entropy* **2020**, *22*, 1347. [CrossRef] [PubMed]
15. Han, T.; Liu, Q.; Zhang, L.; Tan, A. Fault feature extraction of low speed roller bearing based on teager energy operator and ceemd. *Measurement* **2019**, *138*, 400–408. [CrossRef]
16. Li, R.; Ran, C.; Zhang, B.; Han, L.; Feng, S. Rolling bearings fault diagnosis based on improved complete ensemble empirical mode decomposition with adaptive noise, nonlinear entropy, and ensemble svm. *Appl. Sci.* **2020**, *10*, 5542. [CrossRef]
17. Dragomiretskiy, K.; Zosso, D. Variational mode decomposition. *IEEE Trans. Signal Process.* **2014**, *62*, 531–544. [CrossRef]
18. Li, J.; Yao, X.; Wang, H.; Zhang, J. Periodic impulses extraction based on improved adaptive vmd and sparse code shrinkage denoising and its application in rotating machinery fault diagnosis. *Mech. Syst. Signal Process.* **2019**, *126*, 568–589. [CrossRef]
19. Zhang, C.; Yao, W.; Deng, W. Fault diagnosis for rolling bearings using optimized variational mode decomposition and resonance demodulation. *Entropy* **2020**, *22*, 739. [CrossRef]
20. Zhou, X.; Li, Y.; Jiang, L.; Zhou, L. Fault feature extraction for rolling bearings based on parameter-adaptive variational mode decomposition and multi-point optimal minimum entropy deconvolution—Sciencedirect. *Measurement* **2020**, *173*, 108469. [CrossRef]
21. Li, F.; Li, R.; Tian, L.; Chen, L.; Liu, J. Data-driven time-frequency analysis method based on variational mode decomposition and its application to gear fault diagnosis in variable working conditions—Sciencedirect. *Mech. Syst. Signal Process.* **2019**, *116*, 462–479. [CrossRef]
22. Gu, R.; Chen, J.; Hong, R.; Wang, H.; Wu, W. Incipient fault diagnosis of rolling bearings based on adaptive variational mode decomposition and teager energy operator. *Measurement* **2020**, *149*, 106941. [CrossRef]
23. Cai, W.; Yang, Z.; Wang, Z.; Wang, Y. A new compound fault feature extraction method based on multipoint kurtosis and variational mode decomposition. *Entropy* **2018**, *20*, 521. [CrossRef]
24. Li, H.; Liu, T.; Wu, X.; Chen, Q. Application of optimized variational mode decomposition based on kurtosis and resonance frequency in bearing fault feature extraction. *Trans. Inst. Meas. Control* **2019**, *42*, 518–527. [CrossRef]
25. Liu, H.; Xiang, J. Autoregressive model-enhanced variational mode decomposition for mechanical fault detection. *Sci. Meas. Technol.* **2019**, *13*, 843–851. [CrossRef]
26. Miao, Y.; Zhao, M.; Lin, J. Identification of mechanical compound-fault based on the improved parameter-adaptive variational mode decomposition. *ISA Trans.* **2018**, *84*, 82–95. [CrossRef] [PubMed]
27. Liang, T.; Lu, H. A novel method based on multi-island genetic algorithm improved variational mode decomposition and multi-features for fault diagnosis of rolling bearing. *Entropy* **2020**, *22*, 995. [CrossRef] [PubMed]
28. Cheng, C.; Wang, J.; Chen, H.; Chen, Z.; Xie, P. A review of intelligent fault diagnosis for high-speed trains: Qualitative approaches. *Entropy* **2020**, *23*, 1. [CrossRef] [PubMed]

29. Yao, D.; Li, B.; Liu, H.; Yang, J.; Jia, L. Remaining useful life prediction of roller bearings based on improved 1D-CNN and simple recurrent unit. *Measurement* **2021**, *175*, 109166. [CrossRef]
30. Yao, D.; Liu, H.; Yang, J.; Zhang, J. Implementation of a novel algorithm of wheelset and axle box concurrent fault identification based on an efficient neural network with the attention mechanism. *J. Intell. Manuf.* **2020**. [CrossRef]
31. Liu, H.; Yao, D.; Yang, J.; Li, X. Lightweight Convolutional Neural Network and Its Application in Rolling Bearing Fault Diagnosis under Variable Working Conditions. *Sensors* **2019**, *19*, 4827. [CrossRef]
32. Dibaj, A.; Ettefagh, M.M.; Hassannejad, R.; Ehghaghi, M.B. A hybrid fine-tuned vmd and cnn scheme for untrained compound fault diagnosis of rotating machinery with unequal-severity faults. *Expert Syst. Appl.* **2020**, *167*, 114094. [CrossRef]
33. Lu, S.; Sian, H.; Wang, M.; Kuo, C. Fault diagnosis of power capacitors using a convolutional neural network combined with the chaotic synchronisation method and the empirical mode decomposition method. *IET Sci. Meas. Technol.* **2021**. [CrossRef]
34. Rui, Z.; Yan, R.; Chen, Z.; Mao, K.; Gao, R.X. Deep learning and its applications to machine health monitoring. *Mech. Syst. Signal Process.* **2019**, *115*, 213–237.
35. Wen, L.; Gao, L.; Li, X.Y. A new deep transfer learning based on sparse auto-encoder for fault diagnosis. *IEEE Trans. Syst. ManCybern. Syst.* **2017**, *49*, 136–144. [CrossRef]
36. Wang, L.; Liu, Z.; Miao, Q.; Zhang, X. Complete ensemble local mean decomposition with adaptive noise and its application to fault diagnosis for rolling bearings. *Mech. Syst. Signal Process.* **2018**, *106*, 24–39. [CrossRef]
37. Deng, W.; Yao, R.; Zhao, H.; Yang, X.; Li, G. A novel intelligent diagnosis method using optimal ls-svm with improved pso algorithm. *Soft Comput.* **2017**, *23*, 2445–2462. [CrossRef]
38. Li, Y.; Si, S.; Liu, Z.; Liang, X. Review of local mean decomposition and its application in fault diagnosis of rotating machinery. *J. Syst. Eng. Electron.* **2019**, *30*, 799–814.
39. Duan, Y.; Wang, C.; Chen, Y.; Liu, P. Improving the accuracy of fault frequency by means of local mean decomposition and ratio correction method for rolling bearing failure. *Appl. Sci.* **2019**, *9*, 1888. [CrossRef]
40. Wang, J.; Yang, J.; Bai, Y.; Zhao, Y.; Yao, D. A comparative study of the vibration characteristics of railway vehicle axlebox bearings with inner/outer race faults. *Proc. Inst. Mech. Eng. Part F J. Rail Rapid Transit* **2020**. [CrossRef]
41. Deng, M.; Deng, A.; Zhu, J.; Zhai, Y.; Liu, Y. Bandwidth fourier decomposition and its application in incipient fault identification of rolling bearings. *Meas. Sci. Technol.* **2019**, *31*, 015012. [CrossRef]
42. Lei, W.; Liu, Z.; Qiang, M.; Xin, Z. Time–frequency analysis based on ensemble local mean decomposition and fast kurtogram for rotating machinery fault diagnosis. *Mech. Syst. Signal Process.* **2018**, *103*, 60–75. [CrossRef]
43. Dou, C.; Lin, J. Extraction of fault features of machinery based on fourier decomposition method. *IEEE Access* **2019**, *7*, 183468–183478. [CrossRef]
44. Pang, B.; Tang, G.; Tian, T. Enhanced singular spectrum decomposition and its application to rolling bearing fault diagnosis. *IEEE Access* **2019**, *7*, 87769–87782. [CrossRef]
45. Yang, Z.; Boogaard, A.; Chen, R.; Dollevoet, R.; Li, Z. Numerical and experimental study of wheel-rail impact vibration and noise generated at an insulated rail joint. *Int. J. Impact Eng.* **2017**, *113*, 29–39. [CrossRef]
46. Tajalli, M.R.; Zakeri, J.A. Numerical-experimental study of contact-impact forces in the vicinity of a rail breakage. *Eng. Fail. Anal.* **2020**, *115*, 104681. [CrossRef]
47. Choi, J.Y.; Yun, S.W.; Chung, J.S.; Kim, S.H. Comparative study of wheel–rail contact impact force for jointed rail and continuous welded rail on light-rail transit. *Appl. Sci.* **2020**, *10*, 2299. [CrossRef]

Article

A Weighted Subdomain Adaptation Network for Partial Transfer Fault Diagnosis of Rotating Machinery

Sixiang Jia, Jinrui Wang, Xiao Zhang and Baokun Han *

College of Mechanical and Electronic Engineering, Shandong University of Science and Technology, Qingdao 266590, China; sixiang_j@163.com (S.J.); wangjr33@163.com (J.W.); zhangxiao9789@163.com (X.Z.)
* Correspondence: bk_han@163.com; Tel.: +86-139-5325-2858

Abstract: Domain adaptation-based models for fault classification under variable working conditions have become a research focus in recent years. Previous domain adaptation approaches generally assume identical label spaces in the source and target domains, however, such an assumption may be no longer legitimate in a more realistic situation that requires adaptation from a larger and more diverse source domain to a smaller target domain with less number of fault classes. To address the above deficiencies, we propose a partial transfer fault diagnosis model based on a weighted subdomain adaptation network (WSAN) in this paper. Our method pays more attention to the local data distribution while aligning the global distribution. An auxiliary classifier is introduced to obtain the class-level weights of the source samples, so the network can avoid negative transfer caused by unique fault classes in the source domain. Furthermore, a weighted local maximum mean discrepancy (WLMMD) is proposed to capture the fine-grained transferable information and obtain sample-level weights. Finally, relevant distributions of domain-specific layer activations across different domains are aligned. Experimental results show that our method could assign appropriate weights to each source sample and realize efficient partial transfer fault diagnosis.

Keywords: domain adaptation; partial transfer; fault diagnosis; subdomain; rotating machinery

1. Introduction

As indispensable parts of rotating machinery, the fault identification and diagnosis of bearings and gears are crucial for the normal operation of the machinery. Since traditional fault diagnosis methods rely on manual processing of vibration signals, it is difficult to explore the depth of fault diagnosis knowledge. With the widely application in industry and academia of deep learning technology, it is possible to mine effective diagnosis knowledge from massive amounts of fault data [1–3]. Therefore, such methods have been extensively applied in fault diagnosis of rotating machinery [4–7].

Li et al. [8] proposed a fault diagnosis framework based on multi-scale permutation entropy (MPE) and multi-channel fusion convolutional neural networks (MCFCNN). Since it considers the structure and spatial information between different sensor measurement points, the fault diagnosis with high accuracy and speed is realized. Valtierra-Rodriguez et al. [9] proposed a methodology based on convolutional neural networks for automatic detection of broken rotor bars by considering different severity levels. This method applies a notch filter to remove the fundamental frequency component of the current signal, and the short-time Fourier transform (STFT) is used to obtain time-frequency plane. Experimental results show that the methods is capable of identifying the healthy condition of the induction motor. However, the distributions of the collected datasets may different due to the change of the operating environments. The diagnostic knowledge in the original training data will no longer be fully applicable to the new testing data when the working condition changes [10–14]. In this case, the fault diagnosis methods under variable working conditions based on transfer learning come into being. Recently, some transfer learning-based methods have been developed to solve cross domain fault diagnosis problem. Mao et al. [15] proposed a deep dual temporal

domain adaptation (DTDA) model which could recognize whether an early fault occurs and achieve an earlier detection location and lower false alarm rate. An et al. [16] proposed to apply the maximum mean discrepancy (MMD) based on multiple kernels to intelligent fault diagnosis, and the features of different layers were involved in the domain adaptation process. Wang et al. [17] presented a deep adaptive adversarial network (DAAN) which could narrow the discrepancy to learn domain-invariant features. Chen et al. [18] proposed an unsupervised domain adaptation method which could maximize the mutual information between the target feature space and the entire feature space and minimize the feature-level discrepancy between the two domains. Hasan et al. [19] proposed a multitask-aided transfer learning-based diagnostic framework. This method applies multitask learning-based convolutional network to identify working conditions, and then identifies health status of the rolling element bearings based on transfer learning. In a word, transfer learning techniques provide an efficient solution to cross domain fault diagnosis problems.

Although transfer learning-based methods have made great progress, partial transfer fault diagnosis problem has not been well solved. The partial transfer diagnosis means that the number of fault types in the test data is less than that in the training data. Since the machine is in a healthy working state most of the time, the test data may contain only a few types of fault data. That is, the distribution of two domains is different and the label space of target domain is a subset of that of the source domain [20–22]. As many different health types as possible can be involved by training data through a long period of data accumulation, while it is difficult to guarantee the symmetry of health types in testing data and training data. Therefore, this setting is closer to engineering practice compared with the scenario for which the standard domain adaptation is targeted. Since most of the transfer fault diagnosis methods use all source samples for domain adaptation, the unique types of source samples can enable the network to learn false classification knowledge during domain adaptation, which is the major challenge in partial transfer fault diagnosis. Actually, partial transfer problem has been studied in the field of target detection and computer vision. Cao et al. [23] proposed a selective adversarial network (SAN) to facilitate positive transfer by selecting the source samples highly correlated with the target samples. Chen et al. [24] proposed reinforced transfer network (RTNet) which could apply both high-level and pixel-level information to solve partial transfer problem. In addition, importance weighted adversarial nets [25] and example transfer network (ETN) [26] also obtained excellent performance in the image classification task. These works have laid a solid foundation for solving the problem of partial transfer in mechanical fault diagnosis.

Recently, the partial transfer problem has made initial progress in fault diagnosis. Jiao et al. [27] applied weighted cross entropy loss to give smaller weight to the unique source samples, and such weight is determined by the predicted outputs of two classifiers [28]. Li et al. [29] presented a weighted adversarial transfer network (WATN) which used adversarial training to reweight the source domain samples. Yang et al. [30] proposed a deep partial transfer learning network (DPTL-Net) which could learn domain-asymmetry factor to weight the source samples and finally block unnecessary knowledge. The previous partial domain adaptation methods mainly tried to get the weight of the source samples from a global perspective without considering the relationships between two subdomains [31] in source and target domains, which is not conducive to obtaining the fine-grained transferable information in each type of data. To solve the above problem, this paper proposed a weighted subdomain adaptation network (WSAN) to improve the efficiency of partial transfer diagnosis of machinery. All the samples are divided into class-level subdomains, and the subdomain distributions of deep features in multiple layers are aligned. In order to block the samples of outlier source types, an auxiliary classifier is introduced to conduct adversarial training with the feature generator to obtain the class-level weights. To achieve weighted subdomain adaptation, we propose a weighted local maximum mean discrepancy (WLMMD) to measure the Hilbert-Schmidt norm between kernel mean embedding of empirical distributions between relevant subdomains. The main innovations of this work are summarized as follows:

(1) A WSAN framework is presented to solve the partial transfer fault diagnosis problem. Relevant subdomains are built to capture fine-grained transferable information and avoid negative transfer caused by redundant source samples.
(2) The class-level weights are obtained through the adversarial training between the auxiliary classifier and the feature generator. WLMMD is designed to measure the distribution discrepancy between relevant subdomains and obtain fine-grained transferable information. As a result, proper alignment of relevant subdomains in specific activation layers is realized.

The remainder of this work begins with the background of theory in Section 2. In addition, Section 3 provides an introduction to the methodology presented, and Section 4 applies the proposed model to partial transfer fault diagnosis and verifies the advantages of the model by comparing other methods. Finally, some conclusions are drawn in Section 5.

2. Theoretical Background

2.1. Partial Transfer Fault Diagnosis

For standard domain adaptation-based frameworks, target domain D_t and source domain D_s are collected under different but related working conditions [26]. As shown in the upper part of Figure 1, the job of standard transfer fault diagnosis is to facilitate a knowledge transfer from the labeled source data $\{X_s, C_s\}$ to the unlabeled target dataset X_t. However, different from the closed transfer fault diagnosis, the source label space C_s and target label space C_t are different in partial transfer diagnosis problem. In the bottom part of Figure 1, there are more source classes than target classes, i.e., $C_t \subseteq C_s$. In addition, it should be noted that the sample types in the target domain do not deviate from the scope of the source domain, which ensures the authority of the diagnostic knowledge in source domain. The purpose of partial transfer fault diagnosis is to find the categories associated with the source domain and classify them accurately.

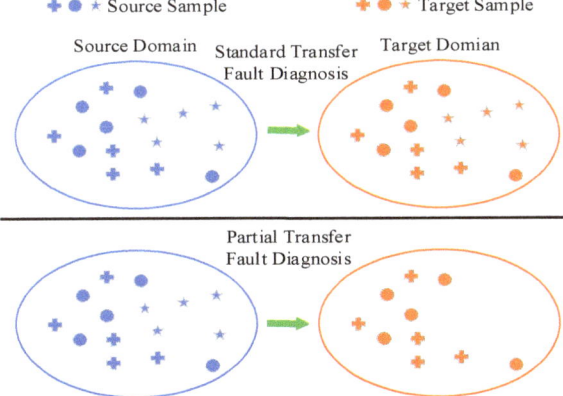

Figure 1. Comparison of standard transfer fault diagnosis and partial transfer fault diagnosis.

2.2. Subdomain Adaptation

The source and target domains may consist of some subdomains that can be defined according to different criteria, such as class or category. For partial transfer fault diagnosis, the number of sample types in the source domain must be no less than that in the target domain, so is practicable to delimit the subdomains based on the number of types in the source domain, although this may not be appropriate for the target domain, but it ensures alignment of local data distribution discrepancy. As can be seen from Figure 2a,b, it is difficult to match two data distributions directly in the process of global or partial domain adaptation. In Figure 2c,d, subdomain adaptation is of superior feature representation ability because the fine-grained transferable information within the subdomains is utilized [31]. However, the problem with this is that the data in the target domain is unlabeled, which prevents target domain from being partitioned. Fortunately, we take the prediction probability output of the model for the target samples as pseudo-labels to divide them into some subdomains. In this way, subdomain adaptation enables the model to focus more on local data distribution differences.

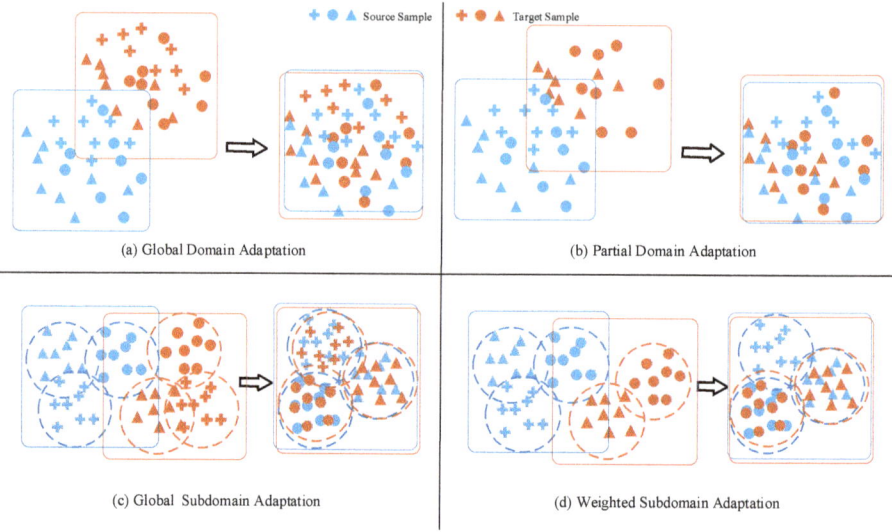

Figure 2. Comparison of standard domain adaptation (**a**,**b**) and subdomain adaptation (**c**,**d**). The box represents the data distribution range and the arrows represent the domain or subdomain adaptation process. The dotted circles represent the divided subdomains.

2.3. Weighted Local Maximum Mean Discrepancy

In the field of transfer learning, MMD [32] is a common nonparametric metric that measures the discrepancy between two distributions. It takes the mean embeddings of two distributions in a Reproducing Kernel Hilbert Space (RKHS) as a distance calculation to avoid the density estimation. MMD can be defined as:

$$d_{\mathcal{H}}(\mathcal{D}_s, \mathcal{D}_t) \triangleq \|\mathbf{E}_p[\phi(x^s)] - \mathbf{E}_q[\phi(x^t)]\|_{\mathcal{H}}^2, \quad (1)$$

where $\phi(\cdot)$ is the feature mapping function that maps the original data to RKHS $\mathcal{H}$. Therefore, an estimate of the MMD compares the square distance between the empirical kernel mean embeddings as:

$$\begin{aligned}\hat{d}_{\mathcal{H}}(\mathcal{D}_s, \mathcal{D}_t) &= \left\|\frac{1}{n_s}\sum_{x_i \in \mathcal{D}_s}\phi(x_i) - \frac{1}{n_t}\sum_{x_j \in \mathcal{D}_t}\phi(x_j)\right\|_{\mathcal{H}}^2 \\ &= \frac{1}{n_s^2}\sum_{i=1}^{n_s}\sum_{j=1}^{n_s}k\left(x_i^s, x_j^s\right) + \frac{1}{n_t^2}\sum_{i=1}^{n_t}\sum_{j=1}^{n_t}k\left(x_i^t, x_j^t\right) \\ &\quad - \frac{2}{n_s n_t}\sum_{i=1}^{n_s}\sum_{j=1}^{n_t}k\left(x_i^s, x_j^t\right)\end{aligned} \quad (2)$$

where $\hat{d}_H(p,q)$ is an unbiased estimator of $d_H(p,q)$. n_s and n_t are the number of source samples and target samples, respectively.

Most previous domain adaptation methods apply MMD to narrow the distribution discrepancy without considering the internal distribution of the data. However, such methods may result in poor alignment because the relationship between related subdomains is ignored. Furthermore, these methods also fail to selectively involve source samples in the adaptation process due to the asymmetry of data types across the two domains. Considering the above problems, we propose the WLMMD to achieve weighted subdomain adaptation:

$$d_{\mathcal{H}}(\mathcal{D}_s, \mathcal{D}_t) \triangleq \mathbf{E}_c \parallel \mathbf{E}_{p^{(c)}}[\phi(x^s)] - \mathbf{E}_{q^{(c)}}[\phi(x^t)] \parallel_{\mathcal{H}}^2, \quad (3)$$

where x^s and x^t are the instances in D_s and D_t, and $p^{(c)}$, and $q^{(c)}$ are the distributions of $D_s^{(c)}$ and $D_t^{(c)}$, respectively. So we can calculate an unbiased estimator of WLMMD as:

$$\hat{d}_{\mathcal{H}}(\mathcal{D}_s, \mathcal{D}_t) = \frac{1}{C} \sum_{k=1}^{C} \parallel \sum_{x_i^s \in \mathcal{D}_s} w_i^{sk} \phi(x_i^s) - \sum_{x_j^t \in \mathcal{D}_t} w_j^{tk} \phi\left(x_j^t\right) \parallel_{\mathcal{H}}^2, \quad (4)$$

where w_i^{sk} and w_j^{tk} denote the weights of x_i^s and x_i^t belonging to class k, respectively. Obviously, $\sum_{i=1}^{n_s} w_i^{sk} = \sum_{i=1}^{n_t} w_i^{tk} = 1$, and w_i^k for the sample x_i can be computed as:

$$w_i^k = \frac{y_{ik}}{\sum_{(x_j, y_j) \in \mathcal{D}} y_{jk}}, \quad (5)$$

where y_{ic} is the k-th entry of vector y_i. Since the source samples are labeled with a one-hot vector, we can directly calculate the weight w_i^{sk} by the labels. Although the samples of the target domain are unlabeled, it is feasible to use pseudo labels to partition related subdomains. Note that the predicted output $\hat{y}_i^t$ given by the classifier can be used as pseudo target labels which measures the probability that the target sample belongs to the corresponding category. $\hat{y}_i^t$ can be regarded as the probability of assigning x_i^t to each of the C classes, and the weight w_i^{tk} of target samples could be acquired. Thus, we can approximate Equation (5) as:

$$\hat{d}_l(\mathcal{D}_s, \mathcal{D}_t) = \frac{1}{C} \sum_{k=1}^{C} \left[\sum_{i=1}^{n_s} \sum_{j=1}^{n_s} w_i^{sk} w_j^{sk} k\left(z_i^{sl}, z_j^{sl}\right) \right.$$
$$+ \sum_{i=1}^{n_t} \sum_{j=1}^{n_t} w_i^{tk} w_j^{tk} k\left(z_i^{tl}, z_j^{tl}\right)$$
$$\left. - 2 \sum_{i=1}^{n_s} \sum_{j=1}^{n_t} w_i^{sk} w_j^{tk} k\left(z_i^{sl}, z_j^{tl}\right) \right] \quad (6)$$

where z^l is the lth layer activation of L layers. By using Equation (6), the distribution discrepancy between the two subdomains at a particular activation layer can be calculated.

3. Proposed Method

3.1. Weighted Subdomain Adaptation Network

In order to achieve efficient partial transfer fault diagnosis, we design a novel weighted subdomain adaptation network (WSAN). The details of the proposed model are clearly presented in Figure 3. The feature generator G is a deep structure based on one dimensional convolutional neural network (1D-CNN) that is expected to extract domain invariant deep features. The auxiliary classifier C_A is set to obtain the class-level weights of the source samples, which is achieved by adversarial training. After acquiring class-level weights, weighted subdomain adaptation can be carried out in activation layers of the classifier C based on WLMMD. The objective function can be written as:

$$\begin{aligned}F_0(\theta_G, \theta_C) &= \frac{1}{n_s}\sum_{i}^{n_s} L_c(G(x_{si}), y_{si})\\ &- \frac{\lambda_0}{n_s+n_t}\sum_{x_i \in (\mathcal{D}_s \cup \mathcal{D}_t)} L_d(D(G(x_i)), d_i)\\ &+ \gamma_0 \hat{d}_l(\mathcal{D}_s, \mathcal{D}_t)\end{aligned} \quad (7)$$

where λ_0 and γ_0 are the penalty coefficients, y_{si} and d_i are the source sample label and domain label, and L denotes the condition prediction loss.

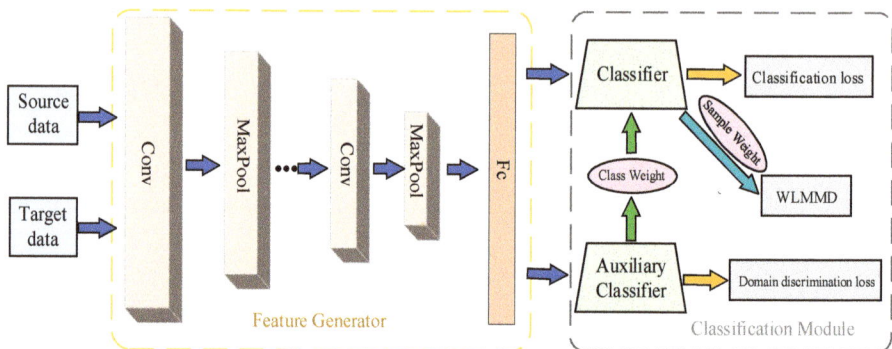

Figure 3. The structural composition of the proposed model.

3.2. Adversarial Training-Based Class-Level Weights Obtaining

Due to the asymmetry of the fault classes in the two domains, samples of redundant types in the source domain may cause a negative transfer. Therefore, these redundant subdomains must be selected to block the classification knowledge that is unfavorable to the recognition of target samples. Inspired by generative adversarial networks (GAN), we set up an auxiliary classifier C_A to play the mini-max game with the feature generator. Specifically, given input x_s or x_t with the label 1 or 0, after multiple layers of extraction, the feature generator G narrows the domain shift to make classifier cannot distinguish the true source of the input sample. The auxiliary classifier is trained to give the correct label. The objective of the adversarial training can be defined as:

$$\begin{aligned}\min_G \max_{C_A} \mathcal{L}(C_A, G) &= \frac{1}{n_s}\sum_{i=1}^{n_s} \log(C_A(G(x_{si})))\\ &+ \frac{1}{n_t}\sum_{j=1}^{n_t} \log(1 - C_A(G(x_{tj})))\end{aligned} \quad (8)$$

The distribution differences of the deep features of shared fault types will be narrowed in the training process, so the auxiliary classifier will be unable to distinguish samples of these types and give an output close to 0, while the output of the unique source samples will be close to 1. The aim of adversarial training is to learn the relative importance of source samples, suggesting that the outlier samples should be assigned a relatively small weight. Therefore, the weight function is inversely related to $C_A(G(x))$ and the importance weights function can be defined as:

$$w_c(x) = 1 - C_A(G(x)), \quad (9)$$

After obtaining the class weights of the source samples, the overall objective can be rewritten as:

$$F(\theta_G, \theta_C) = \frac{1}{n_s} \sum_{i}^{n_s} \sum_{x_{si} \in \mathcal{D}_s^j} w_{cj} L_c(G(x_{si}), y_{si})$$
$$+ \frac{\lambda_1}{n_s} \sum_{i=1}^{n_s} \sum_{x_{si} \in \mathcal{D}_s^j} w_{cj} \log(C_A(G(x_{si})))$$
$$+ \frac{\lambda_2}{n_t} \sum_{j=1}^{n_t} \log\left(1 - C_A\left(G(x_{tj})\right)\right)$$
$$+ \gamma \hat{d}_l(\mathcal{D}_s, \mathcal{D}_t) \tag{10}$$

$$F(\theta_{C_A}) = -\frac{1}{n_s} \sum_{i=1}^{n_s} \sum_{x_{si} \in \mathcal{D}_s^j} w_{cj} \log(C_A(G(x_{si})))$$
$$-\frac{\lambda}{n_t} \sum_{j=1}^{n_t} \log\left(1 - C_A\left(G(x_{tj})\right)\right) \tag{11}$$

where w_{cj} and D_s^j denote the weights and samples for the j-th source class, y_s is the source sample labels and γ is a penalty coefficient.

4. Experiments

4.1. Dataset Introduction

The proposed framework is verified with the datasets collected in our laboratory to validate the performance in partial transfer fault diagnosis. Figure 4a indicates the experimental equipment used in our laboratory. The platform consists of a motor, two balancing rotors, two bearing seats, a planetary gearbox, and a magnetic brake for controlling load. Vibration sensors are installed on fixed holders at both ends of the gearbox, and the sampling frequency is 25.6 kHz.

Figure 4. Representation for rotating machinery fault diagnosis test bed in our laboratory (**a**) and different types of bearing fault (**b**) and gear fault (**c**). Red box indicates the location of damage.

(1) Bearing fault dataset

Five health conditions are involved in the bearing fault dataset, namely, normal, inner ring fault, outer ring fault, rolling element fault, and combined fault of the rolling element and outer ring. The fault parts are shown in Figure 4b. There are two damage sizes for each type of fault, specifically, 0.2 and 0.4 mm. Thus, the bearing fault dataset contains samples of nine health types, namely, NC, IF1, IF2, OF1, OF2, RF1, RF2, RO1, and RO2. Four datasets are obtained under different loads, specifically, L1 (80 N), L2 (60 N), L3 (40 N), and L4 (20 N). And the engine speed is 2000 r/min.

(2) Gear fault dataset

As shown in Figure 4c, the gear fault dataset contains samples of seven health types, namely, normal condition (NC), sun gear fracture (SF), sun gear pitting (SP), sun gear wear (SW), planet gear fracture (PF), planet gear pitting (PP), and planet gear wear (PW). We collected three datasets at different rotational speeds (without load), specifically, S1 (2200 r/min), S2 (2000 r/min), and S3 (1800 r/min).

The number of each type of samples is 500. Thus, the number of samples in bearing and gear datasets are 4500 and 3500, respectively. In order to give full play to the feature extraction and weight learning ability of the proposed method, 40% of the samples were used for training and the remaining for testing.

4.2. Compared Methods

To show the superior performance of the proposed model, four comparative methods are adopted as follows:

(1) Supervised training without classification knowledge transfer is adopted as a basic comparative method (Basic), and it obtained classification knowledge only from the source domain samples.
(2) Domain adaptation framework based on multiple kernel variant of maximum mean discrepancy (MKMMD) [16]: Efficient kernel method is adopted in different layers of the network, and excellent performance was achieved on the global domain adaptation task.
(3) Deep subdomain adaptation network (DSAN) [31]: As a typical global domain adaptation approach, it does not include class-level weight acquisition, that is, the auxiliary classifier is not adopted in the network. In this method, local maximum mean difference (LMMD) is used for effective subdomain adaptation.
(4) Example transfer network (ETN) [26]: It is an adversarial discriminative domain adaptation method, and the adversarial training is adopted to obtain the weights of the source samples. Similar to the proposed method, an auxiliary domain discriminator and an auxiliary classifier are adopted to obtain the sample weights in the source domain.

4.3. Implementation Details

As detailed in Table 1, we randomly discarded a number of fault types to design different partial transfer diagnosis tasks on the basis of the two fault diagnosis datasets. For the dataset, each sample consists of 2400 data points, then fast Fourier transformation (FFT) is applied to transform the time-domain signal to frequency-domain signal that contains 1200 Fourier coefficients. The structure of the framework are illustrated in Table 2. The learning rate is set as 0.0001, and the maximum training epoch is 1000. In order to avoid the effects of random cause, we conducted 10 experiments on each task. The running steps of the proposed model are shown in Algorithm 1. In the test process, the spectral data of the target domain can be directly input into the model for classification. The code programming of the model is implemented on the Pytorch platform.

Algorithm 1: Weighted Subdomain Adaptation Network (WSAN)

Model: Feature generator G; Auxiliary classifier C_A; Classifier C.
Input: Labeled source data $\{X_s, C_s\}$ and unlabeled target data X_t.
For i in **epochs**:
  Step 1: The feature generator G outputs the high-dimensional features of the two domains and inputs them into the feature generator G and classifier C.
  Step 2: Auxiliary classifier C_A obtains the class-level weights. The classifier gives prediction probability output on the target samples and obtains sample-level weights to guide WLMMD to perform subdomain adaptation.
  Step 3: Train the feature generator G and classifier C to obtain the optimal parameters $\hat{\theta}_G$ and $\hat{\theta}_C$ by minimizing $F(\theta_G, \theta_C)$;
  Step 4: Train the auxiliary classifier C_A to obtain the optimal parameters $\hat{\theta}_{C_A}$ by minimizing $F(\theta_{C_A})$;

Table 1. Descriptions of the diagnosis tasks.

Dataset	Task	Transfer	Target	Source
Bearing	B1	L1→L2	All types	All types
	B2	L1→L3	NC, IF1, IF2, OF1, OF2, RF1, RF2, RO2	
	B3	L1→L4	NC, IF1, IF2, OF2, RF1, RO2	
	B4	L2→L3	NC, IF1, OF2, RF1, RO1	
	B5	L2→L1	NC, OF1, OF2, RO1	
	B6	L3→L4	IF1, OF1, RF1	
	B7	L3→L1	IF1, IF2	
	B8	L4→L3	RO2	
Gear	G1	S1→S2	NC, SF, SP, SW, PF, PP	All types
	G2	S1→S3	NC, SF, SP, SW, PF	
	G3	S2→S3	SF, SP, SW, PF	
	G4	S2→S1	SF, SP, SW	
	G5	S3→S1	NC, SP	
	G6	S3→S2	PW	

Table 2. Structural composition of the proposed model.

Module	Layer	Size	Channels × Kernel Size	Stride	Activation
Feature Generator	Input data	1200	/	/	/
	Conv	8 × 300	8 × 3	4	ReLu
	MaxPool	8 × 150	8 × 2	2	/
	Conv	16 × 150	16 × 3	1	ReLu
	MaxPool	16 × 75	16 × 2	2	/
	Conv	32 × 75	32 × 3	1	ReLu
	MaxPool	32 × 37	32 × 2	2	/
	Conv	32 × 37	32 × 2	1	ReLu
	MaxPool	32 × 17	32 × 2	2	/
	Fc	512	/	/	/
Classifier	Fc	256	/	/	/
	Fc	128	/	/	/
	Fc	C	/	/	Softmax
Auxiliary Classifier	Fc	256	/	/	/
	Fc	128	/	/	/
	Fc	1	/	/	Sigmoid

4.4. Experimental Results

As mentioned in Section 2, the deep features in different activation layers of the model are involved in subdomain adaptation. In order to obtain the best performance for subdomain adaptation, the deep features with dimensions 128, 256 and 512 in fully connected layers of the classifier and feature generator (named L1, L2 and L3, respectively) are extracted and combined for comparison. B7 task was selected to verify the combination of different layers, and the experiment was conducted for 15 times. As shown in Figure 5, it is clear that L1 achieves the best performance during the single layer, while L3 performs the worst. It means that the model needs to carry out deep operation to extract more separable domain invariant features. In the multi-layer combination, L1 + L2 performed better than single layer while L1 + L2 + L3 has a lower performance than L1 + L2. This indicates that some non-invariant features may exist in the shallow layers and using subdomain adaptation to align these features will degrade the performance of the model. Therefore, we apply the combination of L1 + L2 for the designed tasks.

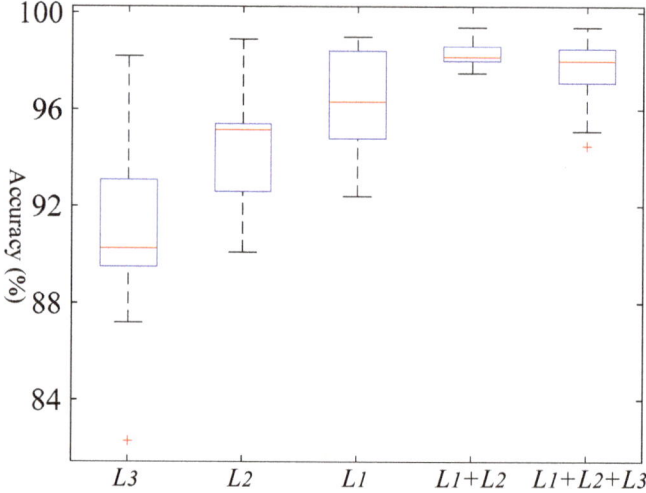

Figure 5. Boxplot for the performance of different layer combinations under the same task.

The average accuracy of the proposed method and the comparison method in all tasks are detailly shown in Table 3. In general, our method obtains the highest average accuracy and the lowest standard deviation. This indicates that WSAN has excellent and stable performance in both global and partial domain adaptation tasks. Since the basic approach does not include any domain adaptation operations, it obtains the worst performance on all tasks. MKMMD achieves the highest accuracy on the non-partial transfer fault diagnosis task B1, but performed poorly on the partial transfer tasks. This indicates that the domain adaptation methods based on MMD has superior performance in the fault diagnosis task under variable conditions, but it is not feasible to directly apply it to partial transfer scenarios. DSAN performed better than MKMMD in most tasks, and its average accuracy is 4.2% higher than that of MKMMD. But it still lags behind the other two partial transfer methods because it does not carry out any weight learning operation. WSAN achieved an average accuracy of 97.7%, which was 4.7% higher than ETN, 13.2% higher than DSAN, and 17.4 higher than MKMMD.

It can be noted that ETN and WSAN, as two domain adaptation methods with weighted learning, perform significantly better than other methods in partial transfer diagnosis tasks. In addition, it can be found that the proposed method gets more ahead of ETN with the increasing degree of domain class asymmetry. For task B2, the accuracy of WSAN is 1.4% higher than that of ETN, while WSAN is 5.6% higher than that of ETN on task B8. The same phenomenon can be observed for tasks G1 and G6.

Table 3. Experimental results of the average testing accuracies in all tasks (%).

Task	Basic	MKMMD	DSAN	ETN	WSAN (Ours)
B1	73.1 (±2.1)	99.5 (±0.1)	98.3 (±0.3)	98.8 (±0.2)	99.4 (±0.1)
B2	65.5 (±4.5)	90.1 (±0.3)	92.6 (±0.8)	97.1 (±0.3)	98.5 (±0.1)
B3	63.7 (±3.4)	87.2 (±0.8)	90.3 (±1.1)	95.5 (±0.5)	98.6 (±0.2)
B4	59.1 (±5.8)	80.4 (±2.1)	90.1 (±1.5)	92.5 (±0.6)	98.5 (±0.2)
B5	59.7 (±4.6)	82.3 (±2.0)	85.4 (±2.6)	93.0 (±0.7)	97.9 (±0.2)
B6	64.0 (±5.8)	79.6 (±3.9)	80.1 (±3.9)	91.4 (±1.0)	98.0 (±0.1)
B7	45.2 (±8.1)	68.2 (±7.3)	75.2 (±5.4)	88.5 (±1.1)	97.8 (±0.3)
B8	42.4 (±6.5)	63.1 (±8.0)	78.9 (±7.2)	92.4 (±0.6)	98.0 (±0.2)
G1	60.2 (±4.4)	88.1 (±0.3)	90.6 (±0.5)	98.4 (±0.6)	99.2 (±0.1)
G2	55.1 (±5.0)	85.2 (±0.8)	90.0 (±1.1)	92.5 (±0.9)	97.1 (±0.1)
G3	50.7 (±5.3)	80.4 (±2.1)	84.1 (±1.5)	94.3 (±1.7)	98.2 (±0.2)
G4	52.0 (±3.8)	82.3 (±2.0)	85.4 (±2.6)	92.4 (±2.0)	95.6 (±0.3)
G5	41.2 (±4.0)	69.6 (±3.9)	71.2 (±3.5)	88.5 (±1.8)	95.1 (±0.3)
G6	35.4 (±5.6)	68.2 (±5.3)	70.2 (±4.0)	88.1 (±0.9)	95.5 (±0.6)
Average	55.2	80.3	84.5	93.0	97.7

To demonstrate the feature classification effect of our method intuitively, the high-dimensional features extracted of the model are processed with the well-known t-SNE [33] technology for dimension reduction. The dimension reduction results of B3 are shown in Figure 6. In Figure 6a, we can see that the feature separability and clustering effect obtained by the basic method are inefficient. Features become separable but shared types and outliers are still cannot be distinguished in Figure 6b,c when domain adaptation is adopted. Although MKMMD and DSAN perform efficient global domain adaptation, the existence of outlier types would enable the model to extract classification knowledge that is not applicable in the target domain. This also indicates that the global adaptation methods only pays attention to the alignment of the two domains, but does not consider the relationship between the subdomains within the domain. In Figure 6d, it can be seen that ETN basically separates outlier samples but the alignment of shared type features is not accurate enough, which indicates that the classifier cannot carry out effective sample-level alignment after obtaining class-level weights and it may leads to inaccurate classification. There are some confusions between the source samples of RO2 and RF1 types. In this case, ETN may treat the RF1 samples as outliers and filter out some useful classification knowledge. For the proposed method, precise alignment of the related subdomains is performed while blocking the outlier types in Figure 6e. After obtaining accurate class-level weights, WSAN can use the proposed WLMMD to perform effective subdomain alignment which involves the sample-level weights learning.

In order to further explore how the weights learned affect the alignment of deep features, the similarity matrix of source and target features in deep layer is drawn on task G4. According to [30], the similarity matrix can be calculated by $G(x_i, x_j) = \exp(-\|x_i - x_j\|^2/200)$ wherein $x_i \in D^s$ and $x_j \in D^t$. Figure 7a shows the actual correspondence between the source and target labels. In Figure 7b, only the samples of SW type can be identified to a certain extent, while the features extracted from the other two target types of samples are highly similar to various source types, which is extremely unfavorable for classification. Obviously, the deep features extracted by the basic method are chaotic due to the lack of domain adaptation operation. In Figure 7c,d, the corresponding samples of SP and SW types have low similarity degree, and some of the samples have great similarity with other types. Consequently, global domain adaptation methods may extract fuzzy deep features when dealing with partial transfer problem. Figure 7e shows that ETN can assign large weight to shared types, but there are still some outlier samples with large weights, resulting in a higher similarity between target features of SP and source features of PP and PW. By comparison, Figure 7f indicates that WSAN obtains more accurate weights, which is reflected in the large similarity between the extracted features of the target domain and corresponding features of source domain, and only a few samples are weakly similar to other source types. In general, the proposed method can make the shared samples fully participate in the subdomain adaptation and block outliers. Thus, the extracted domain invariant features own high similarity among the corresponding shared types.

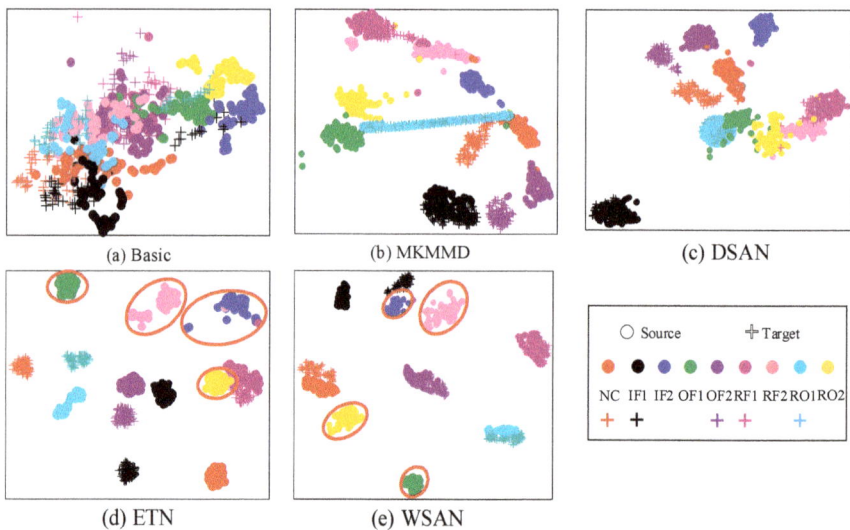

Figure 6. *t*-SNE visualization results of (**a**) Basic, (**b**) MKMMD, (**c**) DSAN, (**d**) ETN, and (**e**) WSAN in task B3. The samples circled in red are outlier types.

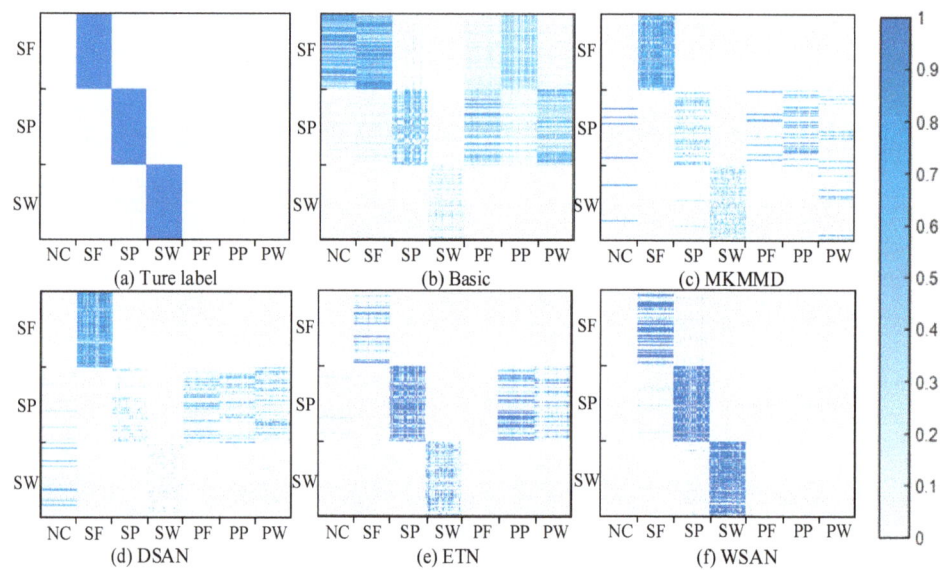

Figure 7. Similarity matrix of learned features by (**a**) ture label, (**b–e**) comparison methods, and (**f**) the proposed method. The abscissa and ordinate represent the source sample sequence and target sample sequence, respectively. The depth of the color indicates the similarity between the corresponding samples.

5. Conclusions

A weighted subdomain adaptation network (WSAN) is presented to solve partial transfer fault diagnosis problem of machinery. Different from the previous global domain adaptation approaches, we divide all samples into different subdomains according to sample types of the source domain, and design WLMMD to perform accurate subdomain alignment. In addition, in order to obtain class-level weights, an additional auxiliary classifier is set up to conduct adversarial training with the feature generator. Under the guidance of class-level weights, the prediction probability output of the target domain by the classifier is used as the sample-level weights, so that the model could capture fine-grained transferable information within the relevant subdomains. The optimal layer combination was found by exploring the performance of the deep features in different activation layers participating the subdomain adaptation. The best diagnostic performance can be obtained under the combination of fully connected layers (L1 + L2) with dimensions 128 and 256. Experimental results on the bearing and gear datasets collected in our laboratory indicates that the average accuracy of the proposed method on the designed fault diagnosis task is 97.7%, which is higher than that of several comparison methods. This means WSAN could solve the partial transfer fault diagnosis problem more efficiently compared several popular methods. *t*-SNE dimension reduction and correlation matrix show that WSAN can learn accurate weights and carry out accurate weighted subdomain adaptation.

Although the proposed weighted subdomain adaptation approach achieves superior performance on the partial transfer fault diagnosis tasks, the laboratory works on the premise that the target data is available during training. It is difficult to guarantee the performance of such a model under unknown working conditions. Such approaches may fail when we need real-time diagnosis. However, this problem may be solved with the help of domain generalization technology [34], and we will explore this issue in depth in our future work.

Author Contributions: Writing—review and editing, S.J.; visualization, J.W.; supervision, X.Z.; project administration, B.H. All authors have read and agreed to the published version of the manuscript.

Funding: The research was supported by the National Natural Science Foundation of China (52005303), the Natural Science Foundation of Shandong Province(ZR201911100329), the Project of China Postdoctoral Science Foundation (2019M662399), and the Postdoctoral Innovation Project of Shandong Province (202003029).

Institutional Review Board Statement: Not applicable.

Informed Consent Statement: Not applicable.

Data Availability Statement: The data presented in this study are available on request from the corresponding author.

Conflicts of Interest: The authors declare no conflict of interest.

References

1. Hoang, D.-T.; Kang, H.-J. A survey on deep learning based bearing fault diagnosis. *Neurocomputing* **2019**, *335*, 327–335. [CrossRef]
2. Li, Y.; Yang, Y.; Li, G.; Xu, M.; Huang, W. A fault diagnosis scheme for planetary gearboxes using modified multi-scale symbolic dynamic entropy and mRMR feature selection. *Mech. Syst. Signal Process.* **2017**, *91*, 295–312. [CrossRef]
3. Lei, Y.; Yang, B.; Jiang, X.; Jia, F.; Li, N.; Nandi, A.K. Applications of machine learning to machine fault diagnosis: A review and roadmap. *Mech. Syst. Signal Process.* **2020**, *138*, 106587. [CrossRef]
4. Shao, H.; Jiang, H.; Zhao, H.; Wang, F. A novel deep autoencoder feature learning method for rotating machinery fault diagnosis. *Mech. Syst. Signal Process.* **2017**, *95*, 187–204. [CrossRef]
5. Wang, J.; Li, S.; An, Z.; Jiang, X.; Qian, W.; Ji, S. Batch-normalized deep neural networks for achieving fast intelligent fault diagnosis of machines. *Neurocomputing* **2019**, *329*, 53–65. [CrossRef]
6. Han, B.; Wang, X.; Ji, S.; Zhang, G.; Jia, S.; He, J. Data-enhanced stacked autoencoders for insufficient fault classification of machinery and its understanding via visualization. *IEEE Access* **2020**, *8*, 67790–67798. [CrossRef]
7. Li, Y.; Wang, X.; Liu, Z.; Liang, X.; Si, S. The entropy algorithm and its variants in the fault diagnosis of rotating machinery: A review. *IEEE Access* **2018**, *6*, 66723–66741. [CrossRef]

8. Li, H.; Huang, J.; Yang, X.; Luo, J.; Zhang, L.; Pang, Y. Fault diagnosis for rotating machinery using multiscale permutation entropy and convolutional neural networks. *Entropy* **2020**, *22*, 851. [CrossRef]
9. Valtierra-Rodriguez, M.; Rivera-Guillen, J.R.; Basurto-Hurtado, J.A.; De-Santiago-Perez, J.J.; Granados-Lieberman, D.; Amezquita-Sanchez, J.P. Convolutional neural network and motor current signature analysis during the transient state for detection of broken rotor bars in induction motors. *Sensors* **2020**, *20*, 3721. [CrossRef]
10. Wang, J.; Feng, W.; Chen, Y.; Yu, H.; Huang, M.; Yu, P.S. Visual domain adaptation with manifold embedded distribution alignment. In Proceedings of the 26th ACM international conference on Multimedia, Seoul, Korea, 22–26 October 2018; pp. 402–410.
11. Zhu, Y.; Zhuang, F.; Wang, J.; Chen, J.; Shi, Z.; Wu, W.; He, Q. Multi-representation adaptation network for cross-domain image classification. *Neural Netw.* **2019**, *119*, 214–221. [CrossRef]
12. Jia, S.; Wang, J.; Han, B.; Zhang, G.; Wang, X.; He, J. A Novel transfer learning method for fault diagnosis using maximum classifier discrepancy with marginal probability distribution adaptation. *IEEE Access* **2020**, *8*, 71475–71485. [CrossRef]
13. Ren, Y.; Li, Y.; Wang, X.; Wang, S.; Si, S. A Novel Bearing Fault Diagnosis Method Based on Multi-scale Transfer Sample Entropy. In Proceedings of the 11th International Conference on Prognostics and System Health Management (PHM-2020 Jinan), Jinan, China, 23–25 October 2020; IEEE: Piscataway, NJ, USA, 2020; pp. 232–236.
14. Wu, Z.; Jiang, H.; Zhao, K.; Li, X. An adaptive deep transfer learning method for bearing fault diagnosis. *Measurement* **2020**, *151*, 107227. [CrossRef]
15. Mao, W.; Sun, B.; Wang, L. A new deep dual temporal domain adaptation method for online detection of bearings early fault. *Entropy* **2021**, *23*, 162. [CrossRef] [PubMed]
16. An, Z.; Li, S.; Wang, J.; Xin, Y.; Xu, K. Generalization of deep neural network for bearing fault diagnosis under different working conditions using multiple kernel method. *Neurocomputing* **2019**, *352*, 42–53. [CrossRef]
17. Wang, J.; Ji, S.; Han, B.; Bao, H.; Jiang, X. Deep adaptive adversarial network-based method for mechanical fault diagnosis under different working conditions. *Complexity* **2020**, *2020*, 6946702. [CrossRef]
18. Chen, J.; Wang, J.; Zhu, J.; Lee, T.H.; De Silva, C. Unsupervised cross-domain fault diagnosis using feature representation alignment networks for rotating machinery. *IEEE/ASME Trans. Mechatron.* **2020**. [CrossRef]
19. Hasan, M.J.; Sohaib, M.; Kim, J.-M. A multitask-aided transfer learning-based diagnostic framework for bearings under inconsistent working conditions. *Sensors* **2020**, *20*, 7205. [CrossRef]
20. Cao, Z.; Ma, L.; Long, M.; Wang, J. Partial adversarial domain adaptation. In Proceedings of the European Conference on Computer Vision (ECCV), Munich, Germany, 8–14 September 2018; pp. 135–150.
21. Li, S.; Liu, C.H.; Lin, Q.; Wen, Q.; Su, L.; Huang, G.; Ding, Z. Deep residual correction network for partial domain adaptation. *IEEE Trans. Pattern Anal. Mach. Intell.* **2020**. [CrossRef]
22. Chen, J.; Wu, X.; Duan, L.; Gao, S. Domain adversarial reinforcement learning for partial domain adaptation. *IEEE Trans. Neural Netw. Learn. Syst.* **2020**. [CrossRef]
23. Cao, Z.; Long, M.; Wang, J.; Jordan, M.I. Partial transfer learning with selective adversarial networks. In Proceedings of the IEEE Conference on Computer Vision and Pattern Recognition, Salt Lake City, UT, USA, 18–22 June 2018; pp. 2724–2732.
24. Chen, Z.; Chen, C.; Cheng, Z.; Jiang, B.; Fang, K.; Jin, X. Selective transfer with reinforced transfer network for partial domain adaptation. In Proceedings of the IEEE/CVF Conference on Computer Vision and Pattern Recognition, Seattle, WA, USA, 14–19 June 2020; pp. 12706–12714.
25. Zhang, J.; Ding, Z.; Li, W.; Ogunbona, P. Importance weighted adversarial nets for partial domain adaptation. In Proceedings of the IEEE Conference on Computer Vision and Pattern Recognition, Salt Lake City, UT, USA, 18–23 June 2018; pp. 8156–8164.
26. Cao, Z.; You, K.; Long, M.; Wang, J.; Yang, Q. Learning to transfer examples for partial domain adaptation. In Proceedings of the IEEE/CVF Conference on Computer Vision and Pattern Recognition, Long Beach, CA, USA, 15–20 June 2019; pp. 2985–2994.
27. Jiao, J.; Zhao, M.; Lin, J.; Ding, C. Classifier inconsistency-based domain adaptation network for partial transfer intelligent diagnosis. *IEEE Trans. Ind. Inform.* **2019**, *16*, 5965–5974. [CrossRef]
28. Saito, K.; Watanabe, K.; Ushiku, Y.; Harada, T. Maximum classifier discrepancy for unsupervised domain adaptation. In Proceedings of the IEEE Conference on Computer Vision and Pattern Recognition, Salt Lake City, UT, USA, 18–23 June 2018; pp. 3723–3732.
29. Li, W.; Chen, Z.; He, G. A novel weighted adversarial transfer network for partial domain fault diagnosis of machinery. *IEEE Trans. Ind. Inform.* **2020**, *17*, 1753–1762. [CrossRef]
30. Yang, B.; Lee, C.-G.; Lei, Y.; Li, N.; Lu, N. Deep partial transfer learning network: A method to selectively transfer diagnostic knowledge across related machines. *Mech. Syst. Signal Process.* **2021**, *156*, 107618. [CrossRef]
31. Zhu, Y.; Zhuang, F.; Wang, J.; Ke, G.; Chen, J.; Bian, J.; Xiong, H.; He, Q. Deep subdomain adaptation network for image classification. *IEEE Trans. Neural Netw. Learn. Syst.* **2020**. [CrossRef]
32. Dziugaite, G.K.; Roy, D.M.; Ghahramani, Z. Training generative neural networks via maximum mean discrepancy optimization. *arXiv* **2015**, arXiv:1505.03906.
33. Van der Maaten, L.; Hinton, G. Visualizing data using t-SNE. *J. Mach. Learn. Res.* **2008**, *9*, 2579–2605.
34. Matsuura, T.; Harada, T. Domain generalization using a mixture of multiple latent domains. In Proceedings of the AAAI Conference on Artificial Intelligence, New York, NY, USA, 7–12 February 2020; pp. 11749–11756.

Article

Fault Detection Based on Multi-Dimensional KDE and Jensen–Shannon Divergence

Juhui Wei [1], Zhangming He [1,2,*], Jiongqi Wang [1], Dayi Wang [2] and Xuanying Zhou [1]

[1] College of Liberal Arts and Sciences, National University of Defense Technology, Changsha 410073, China; weijuhui_nudt@nudt.edu.cn (J.W.); wangjq@nudt.edu.cn (J.W.); zhouxy@nudt.edu.cn (X.Z.)

[2] Beijing Institute of Spacecraft System Engineering, China Academy of Space Technology, Beijing 100094, China; dayiwang@163.com

* Correspondence: hezhangming@nudt.edu.cn; Tel.: +86-1893-248-7430

Abstract: Weak fault signals, high coupling data, and unknown faults commonly exist in fault diagnosis systems, causing low detection and identification performance of fault diagnosis methods based on T^2 statistics or cross entropy. This paper proposes a new fault diagnosis method based on optimal bandwidth kernel density estimation (KDE) and Jensen–Shannon (JS) divergence distribution for improved fault detection performance. KDE addresses weak signal and coupling fault detection, and JS divergence addresses unknown fault detection. Firstly, the formula and algorithm of the optimal bandwidth of multidimensional KDE are presented, and the convergence of the algorithm is proved. Secondly, the difference in JS divergence between the data is obtained based on the optimal KDE and used for fault detection. Finally, the fault diagnosis experiment based on the bearing data from Case Western Reserve University Bearing Data Center is conducted. The results show that for known faults, the proposed method has 10% and 2% higher detection rate than T^2 statistics and the cross entropy method, respectively. For unknown faults, T^2 statistics cannot effectively detect faults, and the proposed method has approximately 15% higher detection rate than the cross entropy method. Thus, the proposed method can effectively improve the fault detection rate.

Keywords: fault detection; optimal bandwidth; kernel density estimation; JS divergence; bearing

1. Introduction

The development of industrial informatization has given rise to a large amount of data in various fields. This has led to data processing becoming a difficult problem in the industry, especially for fault diagnosis. The explosive growth of data provides more information, and therefore, typical data analysis theories often fail in achieving the necessary results. The main reason for this failure can be attributed to the typical data analysis theory that often sets the data distribution type through prior information and performs analyses based on this assumption. Once the distribution type is set, the subsequent analysis can perform the estimation and parametric analysis based on only that distribution type; however, with the growth of data, more information is provided, and thus, the type of data distribution will need to be modified. As a nonparametric estimation method, kernel density estimation (KDE) is the most suitable method for the massive amount of the current data. KDE does not employ a priori assumption for the overall data distribution, and it directly starts from the sample data. When the sample size is sufficient, the KDE can approximate different distributions. Furthermore, Sheather and Jones [1] provides the optimal bandwidth estimation formula for a one-dimensional KDE and proves that the kernel function is asymptotically unbiased and consistent in the density estimation. However, with the growth of the dimension, the multidimensional KDE becomes more complex, and its optimal bandwidth formula is not provided. The distribution of multidimensional data has been described to a certain extent by estimating the kernel density of the reduced data in different dimensions Muir [2], Laurent [3]. In fact, the optimal KDE of multidimensional data is a problem that needs to be studied further.

In the field of fault diagnosis, an essential problem is measuring the difference between samples. A frequency histogram has been used to indicate the distribution difference between two samples Sugumaran and Ramachandran [4], Scott [5]; however, there are three shortcomings to this method: (1) the large number of discrete operations require a higher amount of time; (2) the process depends on the selection of the interval, which is more subjective; (3) there is no intuitive index to reflect this difference. In fact, based on KDE, the JS divergence can be used to measure the difference in data distribution, which can overcome the above shortcomings to a certain extent. For example, the failure of a rolling bearing, which is a key component of mechanical equipment, will have a serious effect on the safe and stable operation of the equipment, and the incipient fault detection of rolling bearings can help avoid equipment running with faults and avoid causing serious safety accidents and economic losses, which has important practical and engineering significance.

In Saruhan et al. [6], vibration analysis of rolling element bearings (REBs) defects is studied. The REBs are the most widely used mechanical parts in rotating machinery under high load and high rotational speeds. In addition, characteristics of bearing faults are analyzed in detail in references Razavi-Far et al. [7], Harmouche et al. [8]. Compared with traditional fault diagnosis, the fault diagnosis of rolling bearings is more complex:

- The fault signal is weak: Bearing data is a type of high-frequency data, and the fault signal is often covered by these high-frequency signals, thereby leading to the failure of traditional fault diagnosis methods. KDE is highly accurate in describing data distribution, so it can identify weak signals.
- Data is highly coupled: Bearing data is reflected in the form of a vibration signal, and there is strong coupling in different dimension signals, thereby making fault diagnosis difficult. Multi-dimensional KDE plays an important role in depicting the correlation of data, which can characterize the relationship between different dimensions of data.
- Incomplete data set: Most bearings work under normal conditions, and the fault data collected are often fewer, which makes the data incomplete, thereby resulting in the imperfection of the fault data set and increasing the difficulty of fault detection. The fault detection method constructed by JS divergence can deal with unknown faults and incomplete data sets without using additional data sets.

To overcome these problems, in-depth research has been conducted on this topic. Fault detection technology based on trend elimination and noise reduction has been proposed previously He et al. [9], Demetriou and Polycarpou [10]. The signal trend ratio is enhanced by eliminating the trend, and the signal–noise ratio is enhanced by noise reduction, and therefore, the fault detection effect is improved. However, this method uses the traditional detection method and cannot effectively solve the problem of data coupling. In reference Zhang et al. [11], Fu et al. [12], a fault detection method based on PCA dimension reduction and modal decomposition feature extraction is proposed. For multidimensional data, PCA dimension reduction is performed to reduce data dimensions and eliminate correlation between different dimensions. Then, the modal decomposition method is used to extract features among dimensions for fault detection. This method can effectively solve the strong coupling between data; however, it will lose some information in the process of PCA dimension reduction, and it leads to a reduction in the fault detection effect. In reference Itani et al. [13], Kong et al. [14], Jones and Sheather [15], Desforges et al. [16], a bearing fault detection method based on KDE is proposed. These studies analyzed the feasibility of KDE method in fault detection, and combined different classification methods for experiments. However, these methods only use one-dimensional KDE, and cannot directly describe high-dimensional data.

The data distribution is reconstructed by KDE and the cross-entropy function is constructed to measure the distribution difference for improving the fault detection results.

However, this method cannot reflect the correlation between different dimensions, and the cross-entropy function is not precise in the description of density distribution, which leads to a reduction in the fault detection effect, especially for unknown fault detection, which is not included in the fault set.

In this study, the KDE method is extended to multidimensional data to avoid information loss caused by the KDE for each dimension, and to better describe the density probability distribution of the data. Meanwhile, this study improves the traditional method using the cross-entropy function as the measurement of density distribution difference, and it uses JS divergence as the measurement of density distribution difference, thereby avoiding the relativity caused by the cross-entropy function. Most fault identification methods are based only on distance measurement; however, only relying on distance measurement cannot effectively detect unknown faults. Based on JS divergence, distribution characteristics of JS divergence between the sample density distribution and population density distribution are derived using the sliding window principle. Thus, the detection threshold of fault identification is assigned to realize the identification of unknown faults.

This paper is based on the following structure. In Section 2, the trend elimination method and detection method are introduced, and the intrinsic and extrinsic signals in the observation data are separated. Then, the fault detection threshold is constructed via statistics. In Section 3, the KDE method is extended to multidimensional data, and the optimal bandwidth is derived. Then, JS divergence is employed to measure the difference between probability distributions of different densities. In Section 4, the sliding window principle is used to sample the training data to obtain the distribution characteristics of JS divergence between the sample density distribution and the overall density distribution, and the detection threshold of fault identification is obtained using the KDE method. In Section 5, the normal data, two known faults, and one unknown fault are identified using the bearing data of the Case Western Reserve University Bearing Data Center as the fault diagnosis data. The experimental results show that the method can identify all types of faults well.

2. T^2 Statistics Fault Detection

In the operation process of the complex equipment or systems, the common observation state can be divided into intrinsic and extrinsic parts. In general, the intrinsic part represents the main working state of the system, which has a certain trend, monotony, and periodicity. The extrinsic part represents system noise, which has a certain zero mean value, high frequency vibration, and statistical stability. For the intrinsic part, the state equation of the system can be used to describe the law. When a fault occurs in the intrinsic part, the symptoms are relatively significant, and the corresponding fault detection methods are relatively mature. However, for high-frequency vibration signals, the incipient fault is often hidden in the extrinsic part, which is easily covered by noise. Therefore, it is necessary to analyze the observed data in depth.

2.1. Signal Decomposition

In the initial operation stage of the equipment, the unstable operation of the system causes large data fluctuations, which will not only have a great effect on the system trend, but also affect the statistical characteristics of the data. Therefore, it is necessary to truncate the data to remove unstable signals [9]. The corresponding time of the time series after removing the nonstationary period data is $t_1, t_2, \cdots, t_m$, and the following m observation data are obtained:

$$Y = [y(t_1), y(t_2), \cdots, y(t_m)]. \quad (1)$$

Each sampling $y(t_i)$ contains n features, which are expressed as components in the form of

$$y(t_i) = [y_1(t_i), y_2(t_i), \cdots, y_n(t_i)]^\mathrm{T}, i = 1, 2, \cdots, m. \quad (2)$$

Then, the data Y can be decomposed into

$$Y = \hat{Y} + R, \quad (3)$$

where $\hat{Y}$ denotes the intrinsic part, which is composed of trend, and R denotes the extrinsic part, which is composed of observation noise and fault data.

The intrinsic part is composed of multiple signals. Selecting the appropriate basis function $f(t) = [f_1(t), f_2(t), \cdots, f_s(t)]^T$ can help describe the intrinsic part. By traversing m data to model the nonlinear data Y,

$$[y_1, y_2, \cdots, y_m] = \begin{bmatrix} \beta_1^{(1)} & \beta_2^{(1)} & \cdots & \beta_s^{(1)} \\ \beta_1^{(2)} & \beta_2^{(2)} & \cdots & \beta_s^{(2)} \\ \vdots & \vdots & \ddots & \vdots \\ \beta_1^{(n)} & \beta_2^{(n)} & \cdots & \beta_s^{(n)} \end{bmatrix} \begin{bmatrix} f_0(t_1) & f_0(t_2) & \cdots & f_0(t_m) \\ f_1(t_1) & f_1(t_2) & \cdots & f_1(t_m) \\ \vdots & \vdots & \ddots & \vdots \\ f_s(t_1) & f_s(t_2) & \cdots & f_s(t_m) \end{bmatrix}. \quad (4)$$

Note that

$$F \triangleq \begin{bmatrix} f_0(t_1) & f_0(t_2) & \cdots & f_0(t_m) \\ f_1(t_1) & f_1(t_2) & \cdots & f_1(t_m) \\ \vdots & \vdots & \ddots & \vdots \\ f_s(t_1) & f_s(t_2) & \cdots & f_s(t_m) \end{bmatrix}, \beta \triangleq \begin{bmatrix} \beta_1^{(1)} & \beta_2^{(1)} & \cdots & \beta_s^{(1)} \\ \beta_1^{(2)} & \beta_2^{(2)} & \cdots & \beta_s^{(2)} \\ \vdots & \vdots & \ddots & \vdots \\ \beta_1^{(n)} & \beta_2^{(n)} & \cdots & \beta_s^{(n)} \end{bmatrix} \quad (5)$$

Then, Equation (4) can be expressed as

$$Y = \beta F. \quad (6)$$

Thus, the efficient estimator of β is

$$\hat{\beta} = YF^T \left(FF^T\right)^{-1}. \quad (7)$$

Using Equations (3) and (7), the signal can be decomposed into

$$\begin{cases} \hat{Y} = \hat{\beta}F = YF^T\left(FF^T\right)^{-1}F \\ R = Y - \hat{Y} = Y\left(I - F^T\left(FF^T\right)^{-1}F\right) \end{cases} \quad (8)$$

Usually, the choice of the basis function is a problem worthy of discussion, and it depends on prior knowledge of practical application scenarios; however, this is not the focus of this paper, and is therefore not covered here.

Remark 1. *For the bearing data, the data is generally stable and periodic. Therefore, Fourier transform is usually used to extract periodic features instead of more complex basis functions, such as a polynomial basis function and wavelet basis function.*

2.2. T^2 Statistics Detection

For simplicity, remember $r_i = r(t_i), i = 1, 2, \cdots, m$. According to Equation (8), the training data after signal decomposition are $R = [r_1, r_2, \cdots, r_m]$, which is generally considered a normal random vector with expectation of $\mathbf{0}$, so that

$$r_i \sim N(\mathbf{0}, \Sigma), \quad (9)$$

where Σ denotes the total covariance matrix. When the covariance matrix Σ is unknown, the unbiased estimation is given by

$$\hat{\Sigma} = \frac{RR^{\mathrm{T}}}{m-1}. \tag{10}$$

Let $Z = [z_1, z_2, \ldots, z_p]$ be the data in the test window to be tested; the sample mean value $\bar{z}$ is

$$\bar{z} = \frac{1}{p}\sum_{i=1}^{p} z_i. \tag{11}$$

Then, $\bar{z}$ is still normal distributed and

$$\bar{z} \sim N\left(0, \frac{1}{p}\Sigma\right). \tag{12}$$

The T^2 statistics can be constructed as

$$T^2 = p\bar{z}^{\mathrm{T}}\hat{\Sigma}^{-1}\bar{z}. \tag{13}$$

Reference Solomons and Hotelling [17] reports that the distribution of the T^2 statistic satisfies

$$\frac{m-n}{n(m-1)}T^2 = \frac{p(m-n)}{n(m-1)}\bar{z}^{\mathrm{T}}\hat{\Sigma}^{-1}\bar{z} \sim F(n, m-n). \tag{14}$$

Therefore, if the significance level is α, we can get that

$$\frac{m-n}{n(m-1)}T^2 = \frac{l(m-n)}{n(m-1)}\bar{z}^{\mathrm{T}}\hat{\Sigma}^{-1}\bar{z} < F_\alpha(n, m-n). \tag{15}$$

The testing data Z and the training data R both come from the same mode; otherwise, they are considered different. The error rate of this criterion is α.

3. Optimal Kernel Density Estimation

Section 2 introduces the fault detection method based on T^2 statistics, including the signal decomposition technology and fault detection method based on the T^2 statistics. However, the fault detection method based on the T^2 statistics assumes that data satisfies the normal distribution, while the actual observation data may not meet the hypothesis, which can lead the discriminant performance of the T^2 statistics to not satisfy the design requirements. In addition, the statistics test the data from the angle of the intrinsic part $\hat{Y}$ and covariance matrix $\hat{\Sigma}$. These two attributes are not sufficient to describe all statistical characteristics of the system. When the incipient fault is submerged by data noise, it is easy to miss the detection. In this study, a KDE method for multidimensional data is constructed to describe the probability and statistical characteristics of the data more accurately.

3.1. Optimal Bandwidth Theorem

For the observed data, the frequency histogram can be used to show its statistical characteristics directly. However, in the actual application process, the frequency histogram is a discrete statistical method, the interval number of the histogram is difficult to divide, and more importantly, the discretization operation inconveniences the subsequent data processing. To overcome these limitations, the KDE method is proposed. This method is a nonparametric estimation method that estimates the population probability density distribution directly by sampling data.

For any point $x \in \mathbb{R}^n$, assuming that the probability density of a certain mode is $f(x)$, the kernel density of $f(x)$ is estimated based on the sampling data $R = [r_1, r_2, \cdots, r_m]$ in Section 2.1. As reported in reference Rao [18], the estimation formula is

$$\hat{f}_K(x, h_m) = \frac{1}{mh_m^n} \sum_{i=1}^m K\left(\frac{r_i - x}{h_m}\right), \tag{16}$$

where m, n, $K(\cdot)$, and h_m denote the number of sampling data, dimension of sampling data, kernel function, and bandwidth, respectively.

For the sake of convenience in the following discussions, in the case of no doubt,

$$\begin{cases} \hat{f}_K(x) \triangleq \hat{f}_K(x, h_m) \\ \int g(x) dx \triangleq \int_{x \in \mathbb{R}^n} g(x) dx \end{cases} \tag{17}$$

The kernel function $K(\cdot)$ satisfies $\int K(x) dx = 1$; therefore, $\int K\left(\frac{r_i - x}{h_m}\right) dx = h_m^n$, that is, $\int \hat{f}_K(x) dx = 1$. Thus, $\hat{f}_K(x)$ satisfies both positive definiteness, continuity, and normality. Therefore, it is reasonable to use it as the KDE. The Gaussian kernel function is a good choice as given by

$$K(x) = (2\pi)^{-n/2} e^{-(x^T x)/2} \tag{18}$$

In this study, the performance of the kernel density estimator is characterized by the mean integral square error (MISE).

$$\text{MISE}\left(\hat{f}_K(x)\right) = \int \text{E}\left[\hat{f}_K(x) - f(x)\right]^2 dx \tag{19}$$

Reference Rao [18] shows that the estimation result $\hat{f}_K(x)$ is not sensitive to the selection of the kernel function $K(\cdot)$; that is, the MISE of the estimation results obtained using different kernel functions is almost the same, which is reflected in the subsequent derivation process. In addition, the MISE depends on the selection of the bandwidth h_m. If h_m is too small, the density estimation $\hat{f}_K(x)$ shows an irregular shape because of the increase in the randomness. While h_m is too large, density estimation $\hat{f}_K(x)$ is too averaged to show sufficient detail.

The optimal bandwidth formula is provided in the following theorem, and it is one of the key theoretical results of this study.

Theorem 1. *For any dimensional probability density function $f(\cdot)$ and any kernel function $K(\cdot)$ with a symmetric form, if $\hat{f}_K(\cdot)$ in Equation (16) is used to estimate $f(\cdot)$, and if the function $\text{tr}\left(\frac{\partial^2 f(x)}{\partial x \partial x^T}\right)$ with respect to x is integrable when the $\text{MISE}\left(\hat{f}_K(\cdot)\right)$ in Equation (19) is the minimum, the bandwidth h_m satisfies*

$$h_m = \left(\frac{md_K^2}{n^3 c_K} \int \text{tr}\left(\frac{\partial^2 f(x)}{\partial x \partial x^T}\right)^2 dx\right)^{-1/(n+4)}, \tag{20}$$

where c_K and d_K are two constant values given by

$$\begin{cases} c_K = \int K^2(x) dx \\ d_K = \int x^T x K^2(x) dx \end{cases} \tag{21}$$

Equation (20) is called the optimal bandwidth formula and h_m denotes the optimal bandwidth.

A detailed proof of this theorem is given below.

Proof. It can be proved that the following two equations hold

$$\begin{cases} \mathrm{E}\left[\hat{f}_K(x)\right] = \int K(u)f(x+h_m u)du \\ \mathrm{E}\left[\hat{f}_K^2(x)\right] = \dfrac{\int K^2(u)f(x+h_m u)du}{mh_m^n} + \dfrac{(m-1)(\int K(u)f(x+h_m u)du)^2}{m} \end{cases} \quad (22)$$

In fact,

$$\begin{aligned} \mathrm{E}\left[\hat{f}_K(x)\right] &= \int \cdots \int \prod_{i=1}^m f(r_i) \frac{1}{mh_m^n} \sum_{i=1}^m K\left(\frac{r_i-x}{h_m}\right) dr_m \cdots dr_1 \\ &= \frac{1}{mh_m^n} \sum_{i=1}^m \int f(r) K\left(\frac{r-x}{h_m}\right) dr \qquad (23) \\ &= \int f(x+h_m u) K(u) du. \end{aligned}$$

In addition,

$$\begin{aligned} \mathrm{E}\left[\hat{f}_K^2(x)\right] &= \int \cdots \int \prod_{i=1}^m f(r_i) \left((mh_m^n)^{-1} \sum_{i=1}^m f(r_i) K\left(\tfrac{r_i-x}{h_m}\right)\right)^2 dr_1 \cdots dr_m \\ &= (mh_m^n)^{-2} \int \cdots \int \prod_{i=1}^m f(r_i) \left(\sum_{i=1}^m f(r_i) K\left(\tfrac{r_i-x}{h_m}\right)\right)^2 dr_1 \cdots dr_m \\ &= (mh_m^n)^{-2} \int \cdots \int \prod_{i=1}^m f(r_i) \left(\sum_{i=1}^m K^2\left(\tfrac{r_i-x}{h_m}\right) + \sum_{i\neq j} K\left(\tfrac{r_i-x}{h_m}\right)K\left(\tfrac{r_j-x}{h_m}\right)\right) dr_1 \cdots dr_m \\ &= (mh_m^n)^{-2} \int \cdots \int \left(\prod_{i=1}^m f(r_i) \sum_{i=1}^m K^2\left(\tfrac{r_i-x}{h_m}\right) + \prod_{i=1}^m f(r_i) \sum_{i\neq j} K\left(\tfrac{r_i-x}{h_m}\right)K\left(\tfrac{r_j-x}{h_m}\right)\right) dr_1 \cdots dr_m \qquad (24) \\ &= (mh_m^n)^{-2} \left(\sum_{i=1}^m \int f(r_i) K^2\left(\tfrac{r_i-x}{h_m}\right) dr + \sum_{i\neq j} \int\int \left(f(r_i) K\left(\tfrac{r_i-x}{h_m}\right) f(r_j) K\left(\tfrac{r_j-x}{h_m}\right)\right) dr_i dr_j\right) \\ &= (mh_m^n)^{-2} \left(m \int f(r) K^2\left(\tfrac{r-x}{h_m}\right) dr + m(m-1) \left(\int f(r) K\left(\tfrac{r-x}{h_m}\right) dr\right)^2\right) \\ &= (mh_m^n)^{-2} \left(mh_m^n \int K^2(u) f(x+h_m u) du + m(m-1)(h_m^n \int f(x+h_m u) K(u) du)^2\right). \end{aligned}$$

From Equation (23),

$$\mathrm{E}\left[\hat{f}_K(x)\right] - f(x) = \frac{h_m^2}{2} \int u^\mathrm{T} \left(\frac{\partial^2 f(x+\theta h_m u)}{\partial x \partial x^\mathrm{T}}\right) u K(u) du, \qquad (25)$$

where θ represents a constant value between 0 and 1. According to Equations (23) and (24),

$$\mathrm{E}\left[\hat{f}_K^2(x)\right] - \left(\mathrm{E}\left[\hat{f}_K(x)\right]\right)^2 = \frac{\int K^2(u) f(x+h_m u) du}{mh_m^n} - \frac{(\int K(u) f(x+h_m u) du)^2}{m}. \qquad (26)$$

According to the Equations (25) and (26), the following equation holds.

$$\begin{aligned} \mathrm{E}\left[\hat{f}_K(x) - f(x)\right]^2 &= \mathrm{E}\left[\hat{f}_K^2(x)\right] - \left(\mathrm{E}\left[\hat{f}_K(x)\right]\right)^2 + \left(\mathrm{E}\left[\hat{f}_K(x)\right] - f(x)\right)^2 \\ &= \frac{\int K^2(u) f(x+h_m u) du}{mh_m^n} - \frac{(\int K(u) f(x+h_m u) du)^2}{m} \qquad (27) \\ &\quad + \left(\frac{1}{2} h_m^2 \int u^\mathrm{T} \left(\frac{\partial^2 f(x+\theta h_m u)}{\partial x \partial x^\mathrm{T}}\right) u K(u) du\right)^2 \end{aligned}$$

To facilitate the subsequent reasoning, the following theorem is given.

Theorem 2. *For any matrix* $\mathbf{\Phi}$, $K(\cdot)$ *is a kernel density function with symmetric form; then,*

$$\int x^T \mathbf{\Phi} x K(x) dx = \frac{tr(\mathbf{\Phi})}{n} \int x^T x K(x) dx. \tag{28}$$

Proof. If the odd function $g(x)$ is integrable on $\mathbb{R}$, there must be $\int_{-\infty}^{\infty} g(x) dx = 0$. Similarly, it can be verified that the kernel function $K(\cdot)$ with a symmetric form satisfies

$$\int \cdots \int \sum_{i \neq j} \Phi_{ij} x_i x_j K(x) dx_1 \cdots dx_n = 0. \tag{29}$$

Then,

$$\int x^T \mathbf{\Phi} x K(x) dx = \int \cdots \int x^T \mathbf{\Phi} x K(x) dx_1 \cdots dx_n$$

$$= \int \cdots \int \sum_i \Phi_{ii} x_i^2 K(x) dx_1 \cdots dx_n + \int \cdots \int \sum_{i \neq j} \Phi_{ij} x_i x_j K(x) dx_1 \cdots dx_n$$

$$= tr(\mathbf{\Phi}) \int \cdots \int x_1^2 K(x) dx_1 \cdots dx_n \tag{30}$$

$$= \frac{tr(\mathbf{\Phi})}{n} \int \cdots \int x^T x K(x) dx_1 \cdots dx_n$$

$$= \frac{tr(\mathbf{\Phi})}{n} \int x^T x K(x) dx.$$

Thus, the Theorem 2 is proved. □

For any unit length vector $u \in \mathbb{R}^n$, the Taylor expansion can be used to obtain

$$\begin{cases} f(x + h_m u) = f(x) + h_m u^T \nabla(f(x)) + o(h_m) \\ \dfrac{\partial^2 f(x + \theta h_m u)}{\partial x_i \partial x_j} = \dfrac{\partial^2 f(x)}{\partial x_i \partial x_j} + \theta h_m u^T \nabla \left(\dfrac{\partial^2 f(x)}{\partial x_i \partial x_j} \right) + o(h_m) \end{cases} \tag{31}$$

If the bandwidth h_m satisfies the condition

$$\begin{cases} \lim_{m \to \infty} (h_m) = 0, \\ \lim_{m \to \infty} \left(\dfrac{1}{m h_m^n} \right) = 0, \end{cases} \tag{32}$$

Then, from Equations (22)–(32), we get that

$$\mathrm{E}\left[\hat{f}_K(x) - f(x)\right]^2 = \frac{c_K f(x)}{m h_m^n} + o\left(\frac{1}{m h_m^n}\right) + \frac{h_m^4 d_K^2}{4n^2} \left(tr\left(\frac{\partial^2 f(x)}{\partial x \partial x^T}\right) \right)^2 + o\left(h_m^4\right). \tag{33}$$

In fact,

$$\mathrm{E}\left[\hat{f}_K(x) - f(x)\right]^2 = \frac{\int K^2(u) f(x + h_m u) du}{m h_m^n} - \frac{\left(\int K(u) f(x + h_m u) du\right)^2}{m}$$

$$+ \left(\frac{h_m^2}{2} \int u^T \left(\frac{\partial^2 f(x + \theta h_m u)}{\partial x \partial x^T} \right) u K(u) du \right)^2$$

$$= \frac{c_K f(x)}{m h_m^n} + o\left(\frac{1}{m h_m^n}\right) - \frac{f(x)^2}{m} + o\left(\frac{1}{m}\right) + \left(\frac{h_m^2}{2n} tr\left(\frac{\partial^2 f(x + \theta h_m u)}{\partial x \partial x^T} \right) \int u^T u K(u) du \right)^2 \tag{34}$$

$$= \frac{c_K f(x)}{m h_m^n} + o\left(\frac{1}{m h_m^n}\right) + \left(\frac{h_m^2}{2n} tr\left(\frac{\partial^2 f(x)}{\partial x \partial x^T} \right) d_K + o\left(h_m^2\right) \right)^2$$

$$= \frac{c_K f(x)}{m h_m^n} + o\left(\frac{1}{m h_m^n}\right) + \frac{h_m^4 d_K^2}{4n^2} \left(tr\left(\frac{\partial^2 f(x)}{\partial x \partial x^T} \right) \right)^2 + o\left(h_m^4\right).$$

Based on Equation (33), if $\text{tr}\left(\frac{\partial^2 f(x)}{\partial x \partial x^T}\right)$ is integrable, there is

$$\text{MISE}\left(\hat{f}_K(x)\right) = \int \left(\frac{c_K f(x)}{mh_m^n} + \frac{h_m^4}{4n^2}\left(d_K \text{tr}\left(\frac{\partial^2 f(x)}{\partial x \partial x^T}\right)\right)^2\right) dx + o\left(\frac{1}{mh_m^n}\right) + o(h_m)$$

$$= \frac{c_K}{mh_m^n} + \frac{1}{4n^2} h_m^4 d_K^2 \int \text{tr}\left(\frac{\partial^2 f(x)}{\partial x \partial x^T}\right)^2 dx + o\left(\frac{1}{mh_m^n}\right) + o(h_m). \quad (35)$$

When $\text{MISE}\left(\hat{f}_K(\cdot)\right)$ is the smallest, the derivative of Equation (35) with respect to h_m is 0, which means

$$\frac{\partial \text{MISE}\left(\hat{f}_K(x)\right)}{\partial h_m} = 0. \quad (36)$$

Thus, the optimal bandwidth h_m in Theorem 1 is obtained as

$$h_m = \left(\frac{m d_K^2}{n^3 c_K} \int \text{tr}\left(\frac{\partial^2 f(x)}{\partial x \partial x^T}\right)^2 dx\right)^{-1/(n+4)}. \quad (37)$$

□

Remark 2. *When the number of samples m is determined, the appropriate bandwidth h_m can be selected using Equation (20) to construct the KDE, which can better fit the sample distribution. In Equation (20), the influence of the kernel function on bandwidth selection is on c_K and d_K, which are almost the same under different kernel function selection, and they have a slight effect on the final bandwidth selection.*

3.2. Optimal Bandwidth Algorithm

The optimal bandwidth formula is given by Equation (20). However, $f(x)$ is unknown in Equation (20), and therefore, $\int \text{tr}\left(\frac{\partial^2 f(x)}{\partial x \partial x^T}\right) dx$ is also unknown. An approximate value of the bandwidth parameter h_m can be obtained by replacing $f(x)$ with $\hat{f}_K(x)$ in Equation (16). Furthermore, an iterative algorithm can be used to calculate a more accurate bandwidth parameter. Theorem 3 shows that the algorithm is convergent.

Theorem 3. *For any n-dimensional probability density function $f(\cdot)$ and Gaussian kernel function $K(\cdot)$, if $\hat{f}_K(\cdot)$ in Equation (16) is used to estimate $f(\cdot)$, then the iterative calculation formula of h_m is obtained as*

$$h_{m,k+1} = \left(\frac{m d_K^2}{n^3 c_K} \int \text{tr}\left(\frac{\partial^2 \hat{f}_K(x, h_{m,k})}{\partial x \partial x^T}\right)^2 dx\right)^{-1/(n+4)} \quad (38)$$

and it is convergent, where $h_{m,k}$ is the value of h_m during the k-th iteration.

Proof. For a particular Gaussian kernel function

$$K(u) = (2\pi)^{-n/2} e^{-\left(u^T u\right)/2} \quad (39)$$

d_K is a χ^2 distribution with degree of freedom n, and the expectation is equal to the degree of freedom.

$$d_K = \int u^T u K(u) du = n \quad (40)$$

In addition,

$$c_K = \int K^2(u) du = \int (2\pi)^{-n} e^{-u^T u} du = (2\sqrt{\pi})^{-n}. \quad (41)$$

Substituting Equations (39)–(40) into Equation (20) and substituting $\hat{f}_K(x)$ in Equation (16) for $f(x)$, the iterative form of calculating h_m is obtained as

$$h_{m,k+1} = \left(\frac{n}{m}\right)^{1/(n+4)} (2\sqrt{\pi})^{-n/(n+4)} \left(\int \text{tr}\left(\frac{\partial^2 \hat{f}_K(x)}{\partial x \partial x^T}\right)^2 dx\right)^{-1/(n+4)}$$

$$= \left(mnh_{m,k}^{2n}\right)^{1/(n+4)} (2\sqrt{\pi})^{-n/(n+4)} \left(\int \text{tr}\left(\frac{\partial^2}{\partial x \partial x^T}\left(\sum_{i=1}^{m} K\left(\frac{r_i-x}{h_{m,k}}\right)\right)\right)^2 dx\right)^{-1/(n+4)} \quad (42)$$

To facilitate the subsequent reasoning, the following lemma is given as

Lemma 1. *For any function $f_1, f_2, \cdots, f_n$, inequality*

$$\int (f_1 + f_2 + \cdots + f_n)^2 dx \leq \int n\left(f_1^2 + f_2^2 + \cdots + f_n^2\right) dx. \quad (43)$$

If and only if $f_1(x) = f_2(x) = \cdots = f_n(x)$ holds almost everywhere.

Proof. In fact, for any function $f_1, f_2, \cdots, f_n$, there are

$$0 \leq (f_1(x) + f_2(x) + \cdots + f_n(x))^2 \leq n\left(f_1(x)^2 + f_2(x)^2 + \cdots + f_n(x)^2\right). \quad (44)$$

Thus, the two sides of Equation (44) are integrated as

$$\int (f_1 + f_2 + \cdots + f_n)^2 dx \leq \int n\left(f_1^2 + f_2^2 + \cdots + f_n^2\right) dx. \quad (45)$$

It is obvious that the sign of Equation (43) holds the condition that $f_1(x) = f_2(x) = \cdots = f_n(x)$ is almost everywhere. □

Because the second derivative of Equation (39) with respect to x_i is

$$\frac{\partial^2}{\partial x_i \partial x_i} K(x) = (2\pi)^{-n/2} e^{-(x^T x)/2} \left(x_i^2 - 1\right). \quad (46)$$

In addition,

$$\int \left(\frac{\partial^2}{\partial x_j \partial x_j}\left(K\left(\frac{r_i-x}{h_{m,k}}\right)\right)\right)^2 dx = \int \left(\frac{\partial^2}{\partial x_j \partial x_j}\left((2\pi)^{-n/2} e^{-(r_i-x)^T(r_i-x)/2h_{m,k}^2}\right)\right)^2 dx$$

$$= \frac{3}{4}(2\sqrt{\pi})^{-n} h_{m,k}^{n-4}. \quad (47)$$

From Lemma 1 and Equation (47)

$$\int \text{tr}\left(\frac{\partial^2}{\partial x \partial x^T}\left(\sum_{i=1}^{m} K\left(\frac{r_i-x}{h_{m,k}}\right)\right)\right)^2 dx \leq \int nm \sum_{i,j}\left(\frac{\partial^2}{\partial x_j \partial x_j}\left(K\left(\frac{r_i-x}{h_{m,k}}\right)\right)\right)^2 dx$$

$$= \frac{3}{4}(nm)^2 (2\sqrt{\pi})^{-n} h_{m,k}^{n-4}. \quad (48)$$

When $h_{m,k}$ is sufficiently large, we can assume that $K\left(\frac{r_i-x}{h_{m,k}}\right)$ is almost the same everywhere, i.e., the equal sign in Equation (48) is tenable.

$$h_{m,k+1} = \left(\frac{mnh_{m,k}^{2n}}{2\sqrt{\pi}}\right)^{1/(n+4)} \left(\frac{3}{4}(nm)^2 (2\sqrt{\pi})^{-n} h_{m,k}^{n-4}\right)^{-1/(n+4)}$$

$$= h_{m,k}\left(\frac{3}{4}nm\right)^{-1/(n+4)} < h_{m,k} \quad (49)$$

When $h_{m,k}$ is large, the iterative process decreases. Because $h_{m,k}$ has a lower bound, the algorithm converges. □

In summary, the KDE method based on optimal bandwidth is given (see Algorithm 1), and the flowchart of the KDE method is shown in Figure 1.

Algorithm 1: Kernel density estimation (KDE) method based on optimal bandwidth

Input: Training set: $R = [r_1, r_2, \cdots, r_m]$; Given the estimation accuracy: ε;
Maximum number of iterations: k_{max}.
Output: Optimal bandwidth: h_m; Optimal KDE: $\hat{f}_K(x)$.

1 Select the initial iteration $h_{m,1} = h_{m,0}$;
2 **for** $k = 1, 2, \cdots, k_{max}$ **do**
3   Calculate the KDE $\hat{f}_K(x)$ using Equation (16)
4   Update the optimal bandwidth $h_{m,k}$ by Equation (38)
5   **if** $k < k_{max} \& |h_{m,k} - h_{m,k-1}| > \varepsilon$ **then**
6          $k = k + 1$, return step3;
7   **else if** $k = k_{max}$ **then**
8   Iteration times overrun, jump out;
9   **else if** $|h_{m,k} - h_{m,k-1}| < \varepsilon$ **then**
10   Obtain the optimal bandwidth.
11   **end**
12 **end**

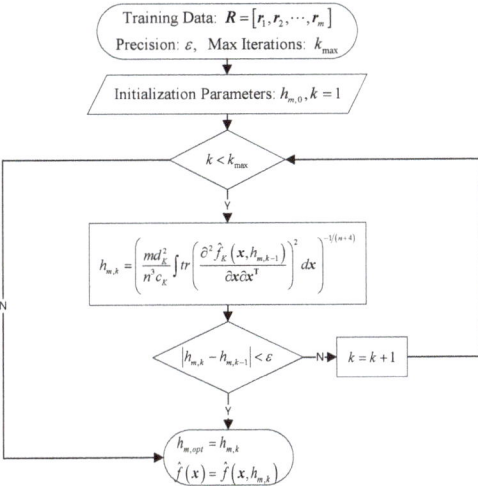

Figure 1. Flowchart of KDE method based on optimal bandwidth.

4. Fault Detection Method Based on JS Divergence Distribution

In Section 3, we construct a multidimensional KDE method based on the optimal bandwidth; this method can accurately describe the density distribution of multidimensional data. JS divergence is used to measure the distribution difference, and thus, it can highlight the difference in the statistical characteristics of different mode data.

4.1. Mode Difference Index

In Section 3, the probability density estimation of multidimensional data is obtained using the kernel function method, and the optimal bandwidth formula is derived. When the system fails, the state of the system will inevitably change, and the statistical characteristics of the system output will also change, thereby leading to significant changes in the density distribution of the observed data. For two groups of the sample window data R and Z, the cross entropy $H(R, Z)$ can be used to measure the distribution difference of R and Z.

$$H(R, Z) = \int -\hat{f}_{K,Z}(x) \log\left(\hat{f}_{K,R}(x)\right) dx, \tag{50}$$

where $\hat{f}_{K,R}, \hat{f}_{K,Z}$ represents the optimal KDE of R and Z calculated using Equation (16).

$H(R, Z)$ does not satisfy the definition of distance because $H(R, Z)$ does not necessarily satisfy positive definiteness and symmetry; that is, $H(R, Z) < 0$ or $H(R, Z) \neq H(R, Z)$.

- The smaller the difference of distribution, the smaller is $H(R, Z)$, which means that even $H(R, Z) < 0$, and therefore, it is reasonable to use $H(R, Z)$ to measure the distribution difference of R and Z.
- However, the quantitative description of distribution difference must satisfy symmetry; otherwise, the exchange position and distribution difference will be different, which is difficult to accept.

The JS divergence $JS(R, Z)$ was used as a measure of the distribution difference between R and Z in reference Zhang et al. [19], Bruni et al. [20] as follows:

$$JS(R, Z) = \int \begin{matrix} \hat{f}_{K,R} \log\left(\hat{f}_{K,R}\right) & +\hat{f}_{K,Z} \log\left(\hat{f}_{K,Z}\right) \\ & -\left(\hat{f}_{K,R} + \hat{f}_{K,Z}\right) \log\left(\left(\hat{f}_{K,R} + \hat{f}_{K,Z}\right)/2\right) \end{matrix} dx. \tag{51}$$

It is easy to get that

$$\begin{cases} JS(R, Z) \geq 0 \\ JS(R, Z) = JS(Z, R) \end{cases} \tag{52}$$

In this paper, Equation (52) is used to measure the distribution difference between testing data Z and training data R for realizing fault detection and isolation.

4.2. Mode Discrimination Method

If the training data has q patterns $\{R_1, R_2, \cdots, R_q\}$, the JS divergence set

$$\{JS(Z, R_1), JS(Z, R_2), \cdots, JS(Z, R_q)\}$$

between the testing data Z and different modes R can be calculated using Equation (51).

If i_0 is the schema tag corresponding to the minimum JS divergence, it means that

$$i_0 = \arg\min\{JS(Z, R_1), JS(Z, R_2), \cdots, JS(Z, R_q)\}. \tag{53}$$

It is reasonable to assume that testing data Z and training data R_{i_0} belong to the same mode. However, for a new failure mode that may be unknown in the application, Equation (50) evaluates the testing data Z as the known failure mode of type i_0, which is obviously unreasonable.

If $JS(Z, R_{i_0})$ is too large, we believe that testing data Z comes from an unknown new failure mode; its label is $q + 1$. However, the method to obtain the threshold JS_{high} of $JS(Z, R_{i_0})$ is a problem that should be investigated. A method to determine JS_{high} is provided below.

For the training data $R_{i_0} = [r_1, r_2, \cdots, r_m]$ of the i_0 mode, the density estimation of the data set can be obtained using Equation (16).

$$\hat{f}_{K,R}(x) = \frac{1}{m(h_m)^n} \sum_{i=1}^{m} K\left(\frac{r_i - x}{h_m}\right) \tag{54}$$

In addition, if the length of the sampling window is fixed as $p (p < m)$, the new sampling data is $\mathbf{R}^{(j)} = [r_j, r_{j+1}, \cdots, r_{j+p}] \subset \mathbf{R}_{i_0}, j = 1, 2, \cdots, m - p$ by sliding the sampling window. For each $\mathbf{R}^{(j)}$, the density of the dataset can be estimated as

$$\hat{f}_{K, \mathbf{R}^{(j)}}(x) = \frac{1}{p(h_p)^n} \sum_{i=j}^{j+p} K\left(\frac{r_i - x}{h_p}\right). \tag{55}$$

Using Equation (52), the divergence between the training data $\mathbf{R}$ and the sample data $\mathbf{R}^{(j)}$ can be obtained as

$$\begin{aligned} JS_j &= JS\left(\mathbf{R}, \mathbf{R}^{(j)}\right) \\ &= H\left(\left(\hat{f}_{K,\mathbf{R}} + \hat{f}_{K,\mathbf{R}^{(j)}}\right), \left(\hat{f}_{K,\mathbf{R}} + \hat{f}_{K,\mathbf{R}^{(j)}}\right)/2\right) - H\left(\hat{f}_{K,\mathbf{R}}\right) - H\left(\hat{f}_{K,\mathbf{R}^{(j)}}\right). \end{aligned} \tag{56}$$

Using Equation (55), we can obtain a series of JS divergence calculation value sets

$$JS = \{JS_1, JS_2, \cdots, JS_{m-p}\}.$$

We use this set to provide the estimation formula $\hat{f}_{JS}(x)$ of the density function $f_{JS}(x)$ of the JS divergence as

$$\hat{f}_{JS}(x) = \frac{1}{(m-p)(h_{m-p})^n} \sum_{j=1}^{m-p} K\left(\frac{JS_j - x}{h_{m-p}}\right). \tag{57}$$

If the significance level is α, the probability of $\hat{f}_{JS}(x)$ that exceeds the threshold JS_{high} is

$$P\left\{\int_0^{JS_{\text{high}}} \hat{f}_{JS}(x) dx.\right\} < \alpha \tag{58}$$

Because the distribution type of JS divergence is not a common random distribution, the quantile cannot be obtained by looking up the table; instead, it can only be obtained by numerical integration. If h is the step size, and

$$\int_{h*(i-1)}^{+\infty} \hat{f}_{JS}(x) dx \leq \alpha \leq \int_{h*i}^{+\infty} \hat{f}_{JS}(x) dx, \tag{59}$$

it is reasonable to deduce that

$$JS_{\text{high}} = h * i. \tag{60}$$

The following fault detection and isolation criteria are constructed by Equation (58).

Criterion 1. *Suppose i_0 is the pattern label corresponding to the minimum JS divergence—see Equation (38)—the training data $\mathbf{R}_{i_0} = [r_1, r_2, \cdots, r_m]$ corresponding to the i_0 mode and the upper bound of JS divergence is JS_{high}—see Equation (56). If the testing data $\mathbf{Z} = [z_1, z_2, \ldots, z_l]$ meet the requirements,*

$$JS(\mathbf{Z}, \mathbf{R}_{i_0}) \leq JS_{\text{high}}. \tag{61}$$

The testing data $\mathbf{Z}$ and training data $\mathbf{R}_{i_0}$ belong to the same failure mode; otherwise, the testing data $\mathbf{Z}$ are considered to originate from the unknown new failure mode, and their label is marked as $q + 1$.

In conclusion, the fault diagnosis method based on optimal bandwidth is provided (See Algorithm 2), and the corresponding fault diagnosis method flowchart is shown in Figure 2.

Algorithm 2: Fault Diagnosis Method Based on Optimal KDE

Input: Training data: $\{R_1, R_2, \cdots, R_p\}$; Significance level: α; Testing data: $Z = [z_1, z_2, \ldots, z_l]$.
Output: Pattern classification labels for testing data Z.

1. Calculate the optimal KDE $JS = \{JS_1, JS_2, \cdots, JS_{m-l}\}$ of R by Algorithm 1;
2. Calculate the optimal KDE $\hat{f}_{K,Z}(x)$ of Z by Algorithm 1;
3. Calculate the JS divergence set $\{JS(Z, R_1), JS(Z, R_2), \cdots, JS(Z, R_p)\}$ of R and Z using Equation (51);
4. Calculate the minimum JS divergence label i_0 using Equation (53), and the corresponding training data were $R_{i_0} = [r_1, r_2, \cdots, r_m] \in R$;
5. **for** $j = 1, 2, \cdots, m - l$ **do**
6.   Get the training data $R^{(j)} = [r_j, r_{j+1}, \cdots, r_{j+l}] \subset R_{i_0}$ by sliding the windows;
7.   Calculate $\hat{f}_K(x)$ based on $h_{m,i}$, kernel function $K(\cdot)$, and Equation (16);
8.   Update the optimal bandwidth $h_{m,i}$ by Equation (37);
9.   Calculate the optimal KDE of $R^{(j)}$ using Algorithm 1 and Equation (55);
10.   Calculate $JS\left(\hat{f}_{K,R}, \hat{f}_{K,R^{(j)}}\right)$ according to Equation (56)
11. **end**
12. Calculate the density function of the JS divergence according to Equation (57) $f_{JS}(x)$;
13. Calculate the upper bound JS_{high} of the JS divergence according to Equation (58) and 60;
14. Assess the pattern of testing data Z according to Criterion 1.

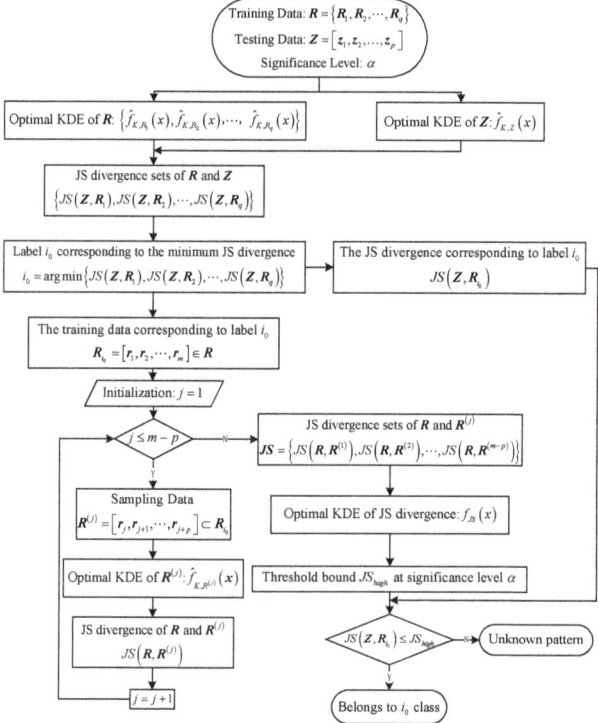

Figure 2. Flowchart of fault diagnosis method based on optimal KDE.

Remark 3. *Equations (54) and (55) show that the calculation result of JS divergence is directly related to the length of sampling data. Indeed, with the increase in the sampling data length, the density estimation obtained by Equation (54) can describe the distribution characteristics of samples more effectively, thereby significantly improving the accuracy of fault detection.*

5. Numerical Simulation

The bearing data from Case Western Reserve University Bearing Data Center were used as the diagnosis research object, and they have been considered as a case for many fault diagnosis, such as in references Smith and Randall [21], Lou and Loparo [22], Rai and Mohanty [23].

The sampling frequency of the motor data was 12 kHz, and 12 kHz is the default sampling frequency for Case Western Reserve University Bearing Data Center. The dataset contains four groups of sample data: normal data (f_0), 0.007 inch inner raceway fault data (f_1), 0.014 inch inner raceway fault data (f_2), and 0.014 inch outer raceway fault data (f_3). Each group of data had two dimensions: the acceleration data of the drive end ($f_i - DE$) and the acceleration data of the fan end ($f_i - FE$). All the experiments were conducted on an Lenovo Ryzen 3700X CPU with 3.60 GHz processor, 16 GB RAM.

5.1. Data Preprocessing

The observed data in the process of the bearing operation show obvious periodicity, which needs to be eliminated. Taking normal data f_0 as an example, the main frequency in the observed signal can be obtained by fast Fourier transform (FFT), and the result of the FFT is shown in Figure 3.

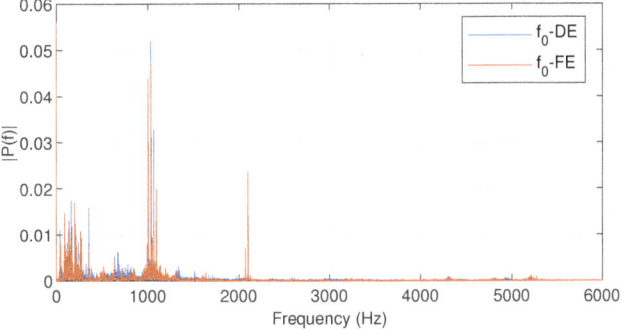

Figure 3. Single-sided amplitude spectrum of f_0.

Figure 3 indicates that the main frequency is approximately 1036 Hz, and thus, the basis function is constructed as

$$f(t) = \begin{bmatrix} 1 & \sin(1036 \times 2\pi t) & \cos(1036 \times 2\pi t) \end{bmatrix}^{\mathrm{T}}.$$

The estimation of β calculated using Equation (7) is

$$\hat{\beta} = \begin{bmatrix} 0.0116 & -0.0158 & 0.0548 \\ 0.0280 & 0.0326 & -0.0396 \end{bmatrix}.$$

Thus, the data after removing the intrinsic signal are shown in Figure 4, where Figure 4a represents the acceleration data of the drive end and Figure 4b represents the acceleration data of the fan end.

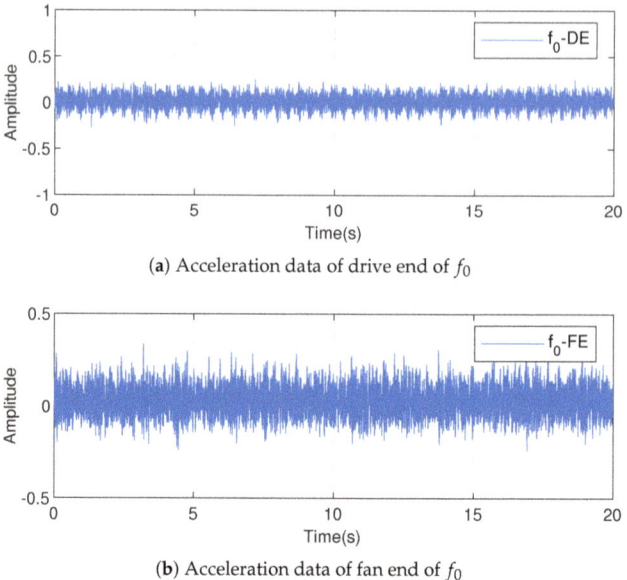

(a) Acceleration data of drive end of f_0

(b) Acceleration data of fan end of f_0

Figure 4. Preprocessed data to remove trends by fast Fourier transform (FFT).

In the later fault detection process, the data of all modes are similar to the above operation, and the results are recorded as f_i.

5.2. Fault Detection Effect
5.2.1. Norm Data and Known Fault

For the norm data f_0 and the known fault f_1, f_2, the first 20,480 sample points are selected as the training set, which are recorded as $f_{i-\text{train}}$. The last 81,920 sample points are taken as the testing set, which are recorded as $f_{i-\text{test}}$. A total of 128 sample points are used as detection objects in each test. The training set data are shown in Figure 5, where Figure 5a,b represent data $f_{i-\text{train}}, i = 1, 2$ of the two dimensions, respectively.

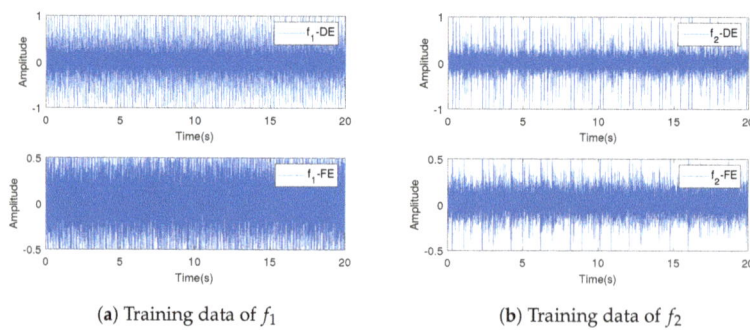

(a) Training data of f_1 (b) Training data of f_2

Figure 5. Training data f_1, f_2 after being preprocessed.

Figure 5 shows that the bearing data have high frequency, and the fault does not change the observed mean value; however, it changes the dispersion characteristics or the correlation of data.

5.2.2. Unknown Fault

The training data does not necessarily contain all types of patterns, and the detection of unknown faults is always a difficult problem. f_3 is used as an unknown fault for fault detection; the training set sample does not contain any information about f_3. The unknown fault data are shown in Figure 6, where in Figure 6a represents the acceleration data at the driving end and Figure 6b represents the acceleration data at the fan end.

Figure 6 shows that the data of unknown faults is close to the other two types of fault data. If the fault detection method is not sensitive, the detection rate will be reduced significantly.

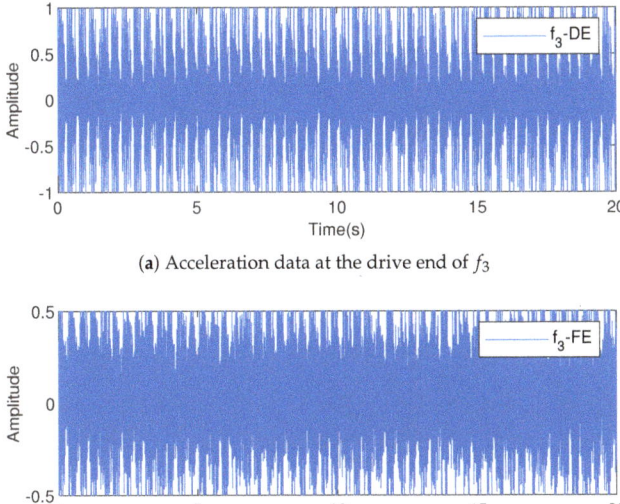

(a) Acceleration data at the drive end of f_3

(b) Acceleration data at the fan end of f_3

Figure 6. Training data f_3 after preprocessed.

5.2.3. Detection Effect

The characteristics of bearing data make bearing fault detection extremely challenging. The input of the training set is $f_{0-train}$, the estimation accuracy is $\varepsilon = 10^{-4}$, and the maximum number of iterations is $k_{max} = 100$, according to Algorithm 1, the optimal bandwidth is

$$h_m = 0.0445.$$

The KDE of the training set is obtained by Equation (15), and the results are shown in Figure 7, where Figure 7a,c,e represent the two-dimensional frequency histograms of the training data $f_{i-train}$, $i = 0, 1, 2$, and Figure 7b,d,f represent the two-dimensional KDE of the training data $f_{i-train}$, $i = 0, 1, 2$.

Figure 7 further shows that the bearing fault changes the dispersion characteristics and data correlation. Meanwhile, Figure 7 shows that the KDE of the training data obtained by Equation (15) is in good agreement with the data distribution of the training data, and therefore, this method can really describe the distribution of multidimensional data.

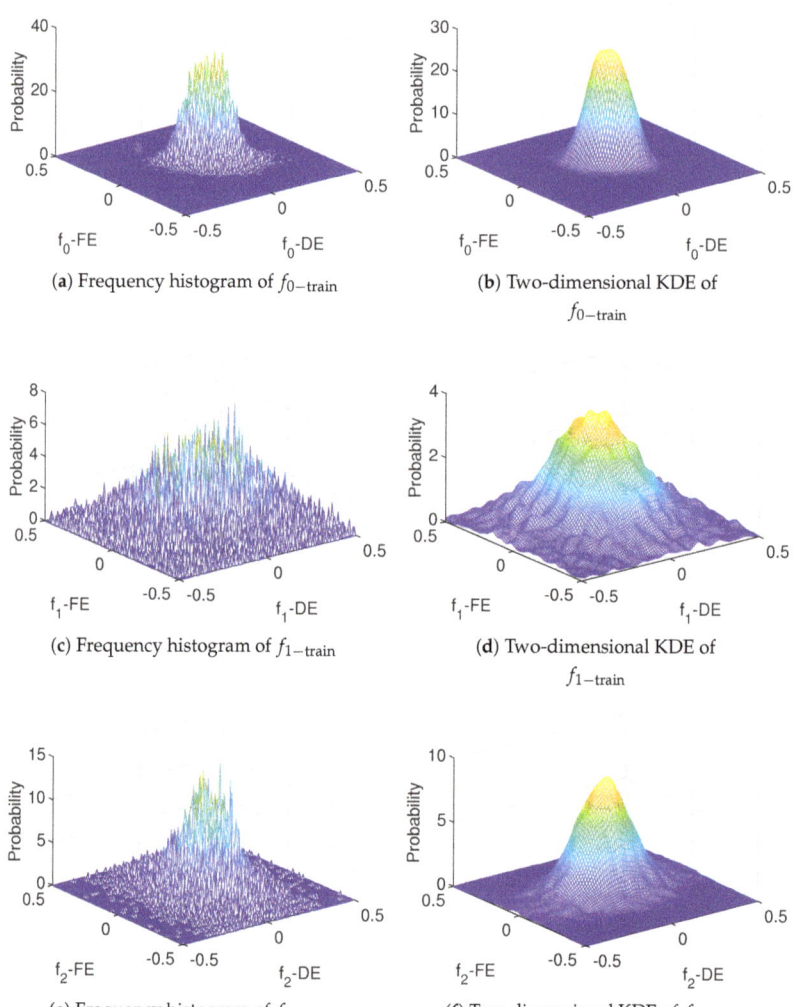

Figure 7. Training data after being preprocessed.

The JS divergence of the training data and KDE of the distribution are obtained by Equations (51) and (58); the results are shown in Figure 8.

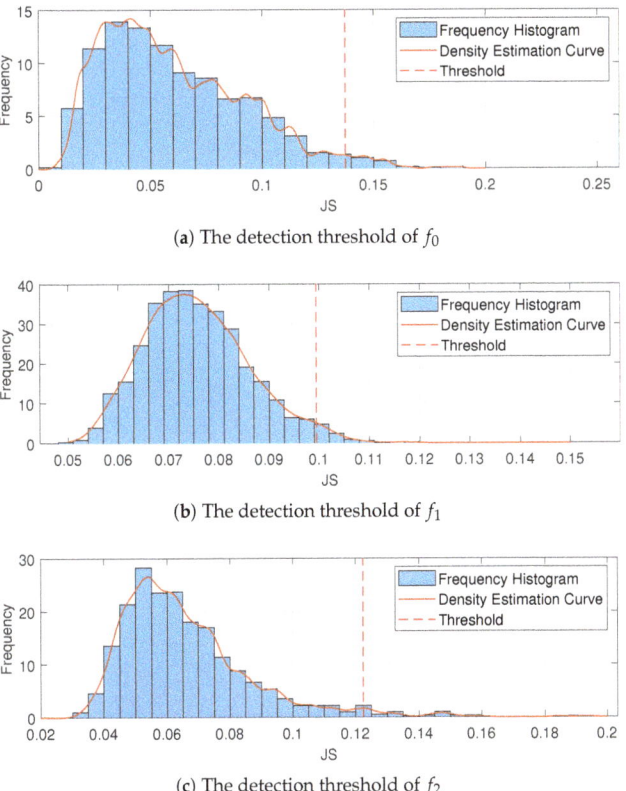

(a) The detection threshold of f_0

(b) The detection threshold of f_1

(c) The detection threshold of f_2

Figure 8. The results of detection threshold.

When the significance level is $\alpha = 95\%$, the detection thresholds of the training set, which are calculated using Equation (58), are

$$\begin{cases} f_0 : JS_{\text{high}} < 0.1375 \\ f_1 : JS_{\text{high}} < 0.0995 \\ f_2 : JS_{\text{high}} < 0.1225 \end{cases}$$

Thus, the detection results of using JS divergence methods on the testing data are shown in Figure 9. If the detection points fall within the threshold, the data set to be detected is in the same pattern; otherwise, the data have different patterns.

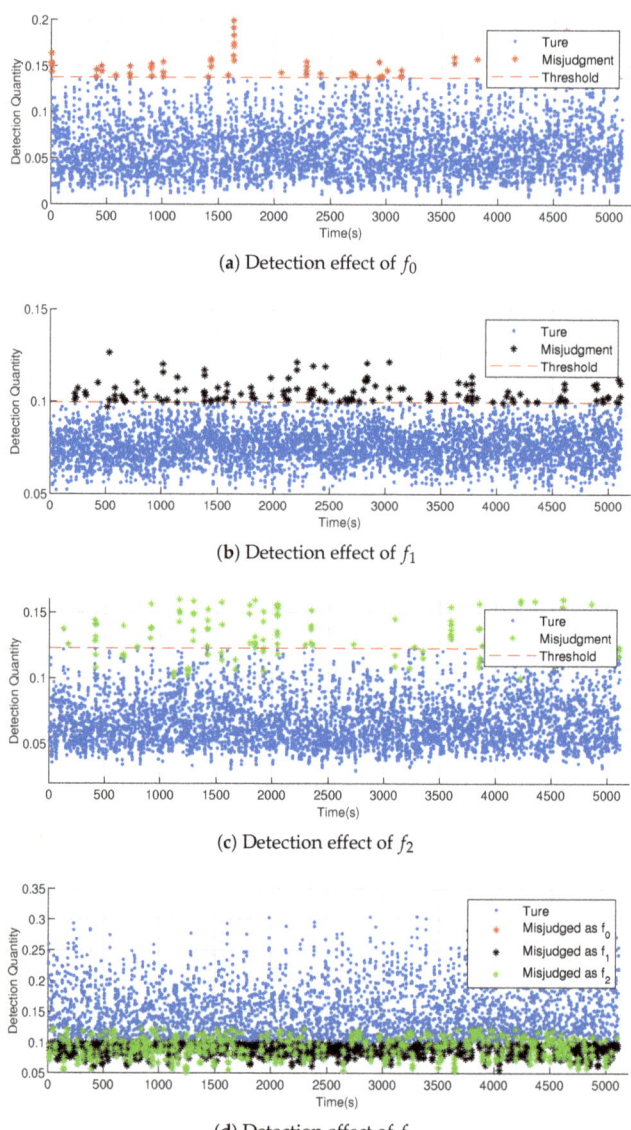

Figure 9. Fault detection effect using JS divergence as index.

Furthermore, detection rates using different methods are shown in Table 1.

Table 1. Detection rate of normal and different failure modes using different methods.

Method	T^2 Statistics Detection	Cross Entropy	JS Divergence
Normal mode f_0	95.80%	96.95%	97.03%
Known fault f_1	83.47%	94.41%	95.81%
Known fault f_2	78.11%	94.19%	95.36%
Unknown fault f_3	\	53.16%	69.49%

For the known fault, Table 1 indicates that the bearing fault identification based on multidimensional KDE and JS divergence achieves better results compared to those obtained using the T^2 statistics detection methods in the testing data. The detection rate of normal data f_0 increases from 95.08% to 97.03%, the detection rate of fault data f_1 increases from 81.33% to 95.81%, and the detection rate of fault data f_2 increases from 70.69% to 95.36%. Meanwhile, compared with the cross-entropy methods, the detection rate of normal data f_0 increased from 96.95% to 97.03%; of fault data f_1 increased from 94.41% to 95.81%; and of fault data f_2 increased from 94.19% to 95.36%.

For the unknown fault f_3, Table 1 shows that the T^2 statistics detection method cannot detect the unknown faults. The method using cross entropy as a measure can only detect unknown faults with a detection rate of 53.16%, which is not obvious. The JS divergence method constructed in this study can identify the unknown fault accurately, and the detection rate reaches 69.49%. This is because JS divergence is more accurate at measuring the differences between distributions.

5.3. Influence of Window Width on Fault Diagnosis

The fault diagnosis effect is related to the data window width; therefore, the fault diagnosis effect under different window widths is investigated. The results are shown in Figure 10.

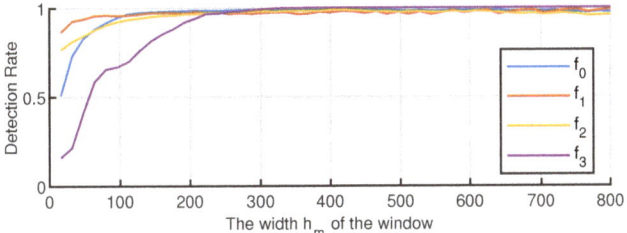

Figure 10. Fault diagnosis effect under different window width h_m.

Figure 10 indicates that, with the increase in the detection window, the detection performance of the proposed method for the known fault detection first rises, and then, it tends to be stable. This is because when the length of the detection window increases to a certain extent, the data to be detected already contains sufficient information. Meanwhile, if the detection window continues to increase, the contribution rate to the improvement of the fault detection rate is not large. Meanwhile, for unknown faults, the detection rate increases rapidly with the length of the detection window because the longer the detection window, the higher the amount of information contained in the data to be detected, and the better is the difference characterized between the fault and the known fault.

6. Conclusions

In this study, a method of bearing fault detection and identification was constructed using multidimensional KDE and JS divergence. The distribution characteristics of JS divergence between the sample density distribution and population density distribution were derived using the sliding sampling window method. Thus, the threshold of fault detection was provided, and therefore, different faults, especially unknown faults, could be identified. The theory showed that the multidimensional KDE method could reduce information loss caused by processing each dimension; the JS divergence is more accurate than the traditional cross entropy to measure the difference in density distribution. The experimental results verified the above conclusions.

For a known fault, the detection effect of this method was obviously better than that of the traditional method, and it also had a certain degree of improvement compared with the cross-entropy method. Second, for unknown faults, the traditional method could not detect the distribution difference accurately, while the detection effect of the proposed method was significantly improved.

Furthermore, the detection effect of this method depends on the window width. The detection effect improved with a growth in the detection window. In this paper, under the condition of a given window width, the estimation formula for the optimal bandwidth of a multidimensional KDE was provided. The experimental results showed that the formula was applicable to any mode of data, and therefore, it had a certain universality.

However, this study has certain limitations. Firstly, although the calculation formula of multidimensional KDE is given in this study, the computational complexity will increase when the dimension is large, which may restrict the further application of the method. Secondly, the calculation of JS divergence is time consuming, which is not conducive to rapid fault diagnosis.

In future research, we can try to use the PCA dimension reduction method to solve the computational complexity caused by very large dimension, and optimize the algorithm flow of JS divergence to expedite the calculation. In the latest study Ginzarly et al. [24], prognosis of the vehicle's electrical machine is treated using a hidden Markov model after modeling the electrical machine using the finite element method. Therefore, we will try to combine this method in future work and apply it to the fault detection of other systems.

Author Contributions: Conceptualization and methodology, J.W. (Juhui Wei); formal analysis and visualization, Z.H.; validation and data curation, J.W. (Jiongqi Wang); resources, D.W.; writing—review and editing X.Z. All authors have read and agreed to the published version of the manuscript.

Funding: This research was funded by the National Natural Science Foundation of China grant number 61903366, 61903086, 61773021, Natural Science Foundation of Hunan Province grant number 2019JJ50745, 2019JJ20018, 2020JJ4280 and Foundation of Beijing Institute of Control Engineering grant number HTKJ2019KL502007.

Institutional Review Board Statement: Not applicable.

Informed Consent Statement: Not applicable.

Data Availability Statement: Not applicable.

Acknowledgments: The authors express their appreciation to the Associate Editor and anonymous reviewers for their helpful suggestions.

Conflicts of Interest: The authors declare no conflict of interest.

Abbreviations

The following abbreviations are used in this manuscript:

KDE	kernel density estimation
JS	Jensen–Shannon
PCA	principal component analysis
MISE	mean integral square error

References

1. Sheather, S.J.; Jones, M.C. A reliable data-based bandwidth selection method for kernel density estimation. *J. R. Stat. Soc.* **1991**, *53*, 683–690. [CrossRef]
2. Muir, D. Multidimensional Kernel Density Estimates over Periodic Domains. Circular Statistics. 2017. Available online: https://www.mathworks.com/matlabcentral/fileexchange/44129-multi-dimensional-kernel-density-estimates-over-periodic-domains (accessed on 21 February 2021).
3. Laurent, B. Efficient estimation of integral functionals of a density. *Ann. Stat.* **1996**, *24*, 659–681. [CrossRef]
4. Sugumaran, V.; Ramachandran, K.I. Fault diagnosis of roller bearing using fuzzy classifier and histogram features with focus on automatic rule learning. *Expert Syst. Appl.* **2011**, *38*, 4901–4907. [CrossRef]
5. Scott, D.W. Averaged shifted histograms: Effective nonparametric density estimators in several dimensions. *Ann. Stat.* **1985**, *13*, 1024–1040. [CrossRef]
6. Saruhan, H.; Sardemir, S.; Iek, A.; Uygur, L. Vibration analysis of rolling element bearings defects. *J. Appl. Res. Technol.* **2014**, *12*, 384–395. [CrossRef]

7. Razavi-Far, R.; Farajzadeh-Zanjani, M.; Saif, M. An integrated class-imbalanced learning scheme for diagnosing bearing defects in induction motors. *IEEE Trans. Ind. Inform.* **2017**, *13*, 2758–2769. [CrossRef]
8. Harmouche, J.; Delpha, C.; Diallo, D. Incipient fault amplitude estimation using kl divergence with a probabilistic approach. *Signal Process.* **2016**, *120*, 1–7. [CrossRef]
9. He, Z.; Shardt, Y.A.W.; Wang, D.; Hou, B.; Zhou, H.; Wang, J. An incipient fault detection approach via detrending and denoising. *Control Eng. Pract.* **2018**, *74*, 1–12. [CrossRef]
10. Demetriou, M.A.; Polycarpou, M.M. Incipient fault diagnosis of dynamical systems using online approximators. *IEEE Trans. Autom. Control* **1998**, *43*, 1612–1617. [CrossRef]
11. Zhang, X.; Polycarpou, M.M.; Parisini, T. A robust detection and isolation scheme for abrupt and incipient faults in nonlinear systems. *IIEEE Trans. Autom. Control* **2002**, *47*, 576–593.
12. Fu, F.; Wang, D.; Li, W.; Li, F. Data-driven fault identifiability analysis for discrete-time dynamic systems. *Int. J. Syst. Sci.* **2020**, *51*, 404–412. [CrossRef]
13. Itani, S.; Lecron, F.; Fortemps, P. A one-class classification decision tree based on kernel density estimation. *Appl. Soft Comput.* **2020**, *91*, 106250. [CrossRef]
14. Kong, Y.; Li, D.; Fan, Y.; Lv, J. Interaction pursuit in high-dimensional multi-response regression via distance correlation. *Ann. Stat.* **2017**, *45*, 897–922. [CrossRef]
15. Jones, M.C.; Sheather, S.J. Using non-stochastic terms to advantage in kernel-based estimation of integrated squared density derivatives. *Stat. Probab. Lett.* **1991**, *11*, 511–514. [CrossRef]
16. Desforges, M.J.; Jacob, P.J.; Ball, A.D. Fault detection in rotating machinery using kernel-based probability density estimation. *Int. J. Syst. Sci.* **2000**, *31*, 1411–1426. [CrossRef]
17. Solomons, L.M.; Hotelling, H. The limits of a measure of skewness. *Ann. Math. Stat.* **1932**, *3*, 141–142.
18. Rao, P. *Nonparametric Functional Estimation*; Elsevier: Amsterdam, The Netherlands, 1983.
19. Zhang, X.; Delpha, C.; Diallo, D. Incipient fault detection and estimation based on Jensen–Shannon divergence in a data-driven approach. *Signal Process.* **2019**, *169*, 107410. [CrossRef]
20. Bruni, V.; Rossi, E.; Vitulano, D. On the equivalence between Jensen–Shannon divergence and Michelson contrast. *IEEE Trans. Inf. Theory* **2012**, *58*, 4278–4288. [CrossRef]
21. Smith, W.A.; Randall, R.B. Rolling element bearing diagnostics using the case western reserve university data: A benchmark study. *Mech. Syst. Signal Process.* **2015**, *64–65*, 100–131. [CrossRef]
22. Lou, X.; Loparo, K.A. Bearing fault diagnosis based on wavelet transform and fuzzy inference. *Mech. Syst. Signal Process.* **2004**, *18*, 1077–1095. [CrossRef]
23. Rai, V.K.; Mohanty, A.R. Bearing fault diagnosis using fft of intrinsic mode functions in Hilbert–Huang transform. *Mech. Syst. Signal Process.* **2007**, *21*, 2607–2615. [CrossRef]
24. Ginzarly, R.; Hoblos, G.; Moubayed, N. From modeling to failure prognosis of permanent magnet synchronous machine. *Appl. Sci.* **2020**, *10*, 691. [CrossRef]

Article

Misalignment Fault Diagnosis for Wind Turbines Based on Information Fusion

Yancai Xiao [1,*], Jinyu Xue [1], Long Zhang [2], Yujia Wang [1] and Mengdi Li [1]

1. School of Mechanical, Electronic and Control Engineering, Beijing Jiaotong University, Beijing 100044, China; 20126082@bjtu.edu.cn (J.X.); shiyanshi10071@163.com (Y.W.); shiyanshi10072@163.com (M.L.)
2. Department of Electrical and Electronic Engineering, The University of Manchester, Manchester M139PL, UK; long.zhang@manchester.ac.uk
* Correspondence: ycxiao@bjtu.edu.cn

Abstract: Most conventional wind turbine fault diagnosis techniques only use a single type of signal as fault feature and their performance could be limited to such signal characteristics. In this paper, multiple types of signals including vibration, temperature, and stator current are used simultaneously for wind turbine misalignment diagnosis. The model is constructed by integrated methods based on Dempster–Shafer (D–S) evidence theory. First, the time domain, frequency domain, and time–frequency domain features of the collected vibration, temperature, and stator current signal are respectively taken as the inputs of the least square support vector machine (LSSVM). Then, the LSSVM outputs the posterior probabilities of the normal, parallel misalignment, angular misalignment, and integrated misalignment of the transmission systems. The posterior probabilities are used as the basic probabilities of the evidence fusion, and the fault diagnosis is completed according to the D–S synthesis and decision rules. Considering the correlation between the inputs, the vibration and current feature vectors' dimensionalities are reduced by t-distributed stochastic neighbor embedding (t-SNE), and the improved artificial bee colony algorithm is used to optimize the parameters of the LSSVM. The results of the simulation and experimental platform demonstrate the accuracy of the proposed model and its superiority compared with other models.

Keywords: wind turbines; misalignment; fault diagnosis; information fusion; improved artificial bee colony algorithm; LSSVM; D–S evidence theory

Citation: Xiao, Y.; Xue, J.; Zhang, L.; Wang, Y.; Li, M. Misalignment Fault Diagnosis for Wind Turbines Based on Information Fusion. *Entropy* **2021**, *23*, 243. https://doi.org/10.3390/e23020243

Academic Editors: Philip Broadbridge, Yongbo Li, Xihui (Larry) Liang and Fengshou Gu

Received: 13 January 2021
Accepted: 18 February 2021
Published: 20 February 2021

Publisher's Note: MDPI stays neutral with regard to jurisdictional claims in published maps and institutional affiliations.

Copyright: © 2021 by the authors. Licensee MDPI, Basel, Switzerland. This article is an open access article distributed under the terms and conditions of the Creative Commons Attribution (CC BY) license (https://creativecommons.org/licenses/by/4.0/).

1. Introduction

In order to address global warming issues, many countries have reduced carbon emissions year by year as one of their targets for economic and social development. As one typical source of clean energy, wind power has significant advantages in terms of environmental and ecological impact compared with hydropower and nuclear power [1]. In recent years, wind power has been rapidly developed in many countries, and the installed capacity has been increasing year by year [2].

The working environment of wind turbines is often complex, so the failure rate of the components of wind turbines is relatively high [3]. If the key components of the wind turbine system fail, it will cause damage and even stop the whole turbine, resulting in huge economic losses. Therefore, in recent years, a large number of research work has been focused on fault diagnosis of wind turbines. The failures typically include blade failures, transmission system failures, generator failures, and tower failures. Among them, misalignment of the transmission system is one of the common failures [4]. Many reasons, such as bearing eccentricity, installation error, and coupling misalignment, can cause misalignment of the wind turbine transmission system that connects the gearbox and generator for a typical doubly-fed wind turbine [5]. The misalignment of the transmission system can inevitably lead to vibration of the unit, which will reduce the reliability of the power generation system. In addition, the misalignment failure can cause damage to gears and bearings [6]. Therefore, it is necessary to monitor and diagnose the misalignment of the transmission system in doubly-fed wind turbines.

Although there is much work on the misalignment fault diagnosis for a conventional rotating system, there is little work for wind turbine misalignment diagnosis. In particular, a wind turbine presents additional and unique challenges as it operates under variable rotational conditions [7,8]. At present, the main research on detecting the misalignment of wind turbines includes the following work. Zhao et al. applied variational mode decomposition (VMD) to decompose the fault vibration signal to isolate features and diagnose the misalignment faults in a direct drive wind turbine [9]. Abdalla et al. diagnosed misalignment of planetary gearbox based on vibration measurements using spectrum analysis and modulation signal bispectrum (MSB) analysis [10]. Huang et al. applied the Hilbert–Huang transform (HHT) method for fault diagnosis of wind turbine rotors and discussed three typical faults by the HHT, including rotor mass imbalance, aerodynamic asymmetries, and yaw misalignment [11]. An and Kong proposed a modified empirical mode decomposition (EMD) method to extract characteristics from vibration signals and applied a back-propagation neural network to data from various sensors to diagnose faults of offshore wind turbines included stator imbalanced, rotor unbalanced, and bearing misalignment [12]. Villa et al. developed a statistical diagnosis algorithm based on the significance level of the modeled fault to detected unbalance fault and misalignment fault of wind turbine, and tested the algorithm on vibration from a test-bed [13]. He et al. analyzed the vibration characteristics of the transmission chain of a wind turbine based on double-elastic support with natural axial misalignment between the output shaft of gearbox and the shaft of generator causing vibration signals of normal gearbox blend with serious high-order gear mesh frequency and smooth modulation [14]. However, these methods mainly applied rely on single information, and their performance could be limited owing to the limited source of information.

Because the diagnosis based on single information often cannot reflect the overall condition, the information fusion methodology for multiple source information is needed for the diagnostic system. Information fusion is a synchronous and comprehensive processing of the information obtained from multiple sensors. It can ensure the integrity of the information from a different perspective and overcome the shortcomings of traditional single information to form a more objective and closer understanding of the system [15], which can greatly improve the accuracy of diagnosis.

Information fusion can be divided into three levels: data level, feature level, and decision level [16,17].

- Data level fusion. The direct fusion of signals collected by the same type of sensors retains the most information among the three levels.
- Feature level fusion. In this process, the signals from multiple sensors need to be preprocessed. Features are extracted to form the fusion vector and its attributes are used to judge the state of targets to be diagnosed.
- Decision level fusion. After initial state judgment of the target to be diagnosed, the final state is obtained based on the fusion of some decision rules. Decision level fusion is the highest among the three levels. Its real-time performance and fault tolerance are very good, but the information loss is very large, so more complex algorithms are needed.

At present, there are many research methods and achievements in decision level fusion, including Bayesian theory [18], Dempster–Shafer (D–S) evidence theory [19], fuzzy set theory [20,21], rough set theory [22], and so on. The classification principle of Bayesian theory is to calculate the posterior probability of an object (the probability that the object belongs to a certain class) using the prior probability and Bayes formula, and select the class with the largest posterior probability as the one to which the object belongs. In D–S evidence theory, trust function and likelihood function are obtained by calculating the orthogonal sum of basic probability distribution functions of different evidences. After fusing multiple evidences, the final decision is made according to decision rules. Among them, basic probability distribution function is the probability distribution of all possible faults in each state, trust function is the lower bound of fault event probability, and likelihood function is the upper bound. Fuzzy set theory (FS) was founded by Zadeh. Membership T(x) was used to describe fuzzy information. At this time, non-membership F(x) did not appear. Then, intuitionistic fuzzy sets (IFSs) and interval intuitionistic fuzzy sets (IVIFSs) appeared successively. The fuzzy information processing technology developed from fuzzy set theory can provide a simple and effective means to explore uncertainty and simulate human recognition mechanism. Rough set theory, initially developed by Pawlak (1982), is a mathematical tool that deals with vague, uncertain, and incomplete information. Rough set theory has been successfully applied in many fields such as machine learning, pattern recognition, control systems, data mining, and image classification.

The advantages and limitations of the above four methods are listed in Table 1.

Table 1. Comparison of information fusion algorithms. D–S, Dempster–Shafer.

Approach	Advantages	Disadvantages
Bayes' theorem	Takes probability as the input data, has sufficient theoretical knowledge	Difficult to define prior probability function, lacks the ability to allocate the total uncertainty
D–S evidence theory	The premise is easier to meet, no need to know prior probability	Cannot solve the serious conflict or complete conflict of evidences
Fuzzy set theory	Based on local theory of classification, has strong adaptive ability	Determines the uncertainty according to the subjective judgement
Rough set theory	Deals with redundant information and inconsistent information effectively	Discretization of symptom attributes is needed

In this paper, based on the good theoretical basis and application effect of D–S evidence theory [23–27], it is used to complete decision fusion, which provides a sufficient fault diagnosis solution for wind turbine misalignment fault.

The aim of this paper is to use multiple sources of information to distinguish the misalignment-free (normal condition) and three different types of transmission misalignment. The main contributions are summarized as follows.

Multiple sources of information and integrated approach are used for wind turbine transmission misalignment detection. More specifically, the vibration, temperature, and stator current signal are taken as the original source, and their time domain features, frequency domain features, and time-frequency domain features are extracted as fault characteristics. t-distributed stochastic neighbor embedding (t-SNE) is used to reduce the vibration and current characteristics dimensionality, and then three posterior probability least squares support vector machine with parameters optimized by improved artificial bee colony algorithm are constructed. The probability outputs of the three LSSVM are taken as the basic probabilities of evidence fusion. The probability distribution after fusion is calculated according to the Dempster fusion rule. Compared with the non-fusion models, it is demonstrated that the model based on D–S evidence fusion has higher diagnostic accuracy for wind turbine misalignment faults.

The remainder of the paper is organized in the following way. In Section 2, the formulas of D–S evidence theory, posterior probability least squares support vector machines, and the improved artificial bee colony are presented in detail. Section 3 describes the specific steps for D–S fault diagnosis. Section 4 presents the fault diagnosis case study based on the simulation model. Section 5 presents the fault diagnosis case study based on the experimental platform. Section 6 concludes the current work.

2. Theoretical Background

2.1. D–S Evidence Theory

The D–S evidence theory is a method of uncertainty reasoning, proposed by Dempster in 1967 and later improved and developed by Shafer [28]. The D–S evidence method can produce a probability interval to an uncertain event by fusing multiple evidences with known probability distribution. As an indeterminate reasoning method, D–S evidence theory uses weaker conditions than Bayesian, and has the ability to quantify unknown and uncertainty [29]. The evidence theory contains three important functions: basic probability assignment function, belief function, and plausibility function. The basic probability assignment function is the probability distribution of all possible faults in each state, the belief function is the lower bound of the probability of the fault event, and the plausibility function is the upper bound of the probability of the fault event. The belief function and the plausibility function can be obtained by calculating the sum of the basic probability assignment function, and the final decision is made after combining multiple evidences from different sources.

The D–S evidence theory consists of the following parts [30].

- Frame of discernment:

A variety of possible mutually exclusive hypothesis $X_i (i = 1, 2, \cdots, s)$ of a question constitute a finite and non-empty set, which is called the frame of discernment, denoted as $\Omega = \{X_1, X_2, \cdots, X_s\}$.

- Basic probability assignment (BPA) function:

BPA function is also known as the mass function. Suppose H is a subset of Ω, if function $m(H)$ satisfying

(a) $m(\phi) = 0$
(b) $\sum m(H) = 1$
(c) $m(H) > 0$

then, function $m(H)$ is called the basic probability assignment of H on Ω.

- Belief function:

In the frame of discernment, the belief function represents the sum of the basic probability assignment functions of all subsets of H. The expression of the belief function is as follows:

$$bel(H) = \sum_{Y/Y \subseteq H} m(Y) \tag{1}$$

- Plausibility function:

In the frame of discernment, the plausibility function represents the degrees of belief for not denying H, which is the sum of the basic probability assignments of all the subsets intersecting H. The expression of the plausibility function is as follows:

$$pl(H) = \sum_{S/S \cap H \neq \emptyset} m(S) \tag{2}$$

- Dempster's rule of combination:

Dempster's rule of combination is used to combine the BPA functions of multiple evidences. Although this rule is controversial at present, the authors of [31] have showed that it behaves perfectly when evidences do not conflict reciprocally. Only if we integrate conflicting evidences do we need to improve it. In this paper, there is no serious and complete conflict among the outputs from vibration signal, temperature signal, and stator current signal as evidences in this study. Therefore, Dempster's rule is still used here.

Suppose there are n independent evidences (sensors or expert opinions), $H_1, H_2, \cdots, H_n$ (are subsets of Ω), the BPA of them are $m_1, m_2, \cdots, m_n$. Then, Dempster's rule for the BPA functions on Ω is as follows:

$$m = m_1 \oplus m_2 \oplus \cdots \oplus m_n \qquad (3)$$

Specifically, it can be expressed as follows:

$$m(H) = \frac{\sum_{H_1 \cap \cdots \cap H_n} m_1(H_1) m_2(H_2) \cdots m_n(H_n)}{1 - K}, H \neq \varnothing \qquad (4)$$

where the expression of K is as follows:

$$K = \sum_{H_1 \cap \cdots \cap H_n = \varnothing} m_1(H_1) m_2(H_2) \cdots m_n(H_n) \qquad (5)$$

where K is the degree of conflict between evidences. When $K = 1$, the evidences are completely conflicted and cannot be synthesized by this formula; when K tends to 1, the evidences are highly conflicted, and synthesizing by this formula may lead to results contrary to fact [32].

- Decision rules:

The decision rule is to draw a diagnosis based on the uncertain interval $[bel(H), pl(H)]$ of the evidence. In the interval of $[0,1]$, the uncertainty of a proposition is shown in Figure 1.

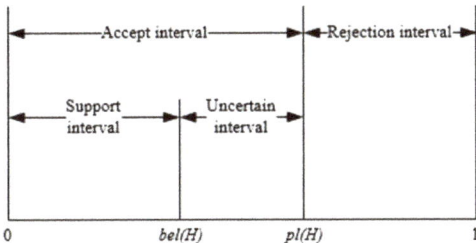

Figure 1. Uncertainty representation of a proposition.

In Figure 1, $[0, bel(H)]$ belongs to the support interval, $[0, pl(H)]$ is the accept interval, $[pl(H), 1]$ is the rejection interval, and $[bel(H), pl(H)]$ is the uncertain interval.

When making a decision, choose a value in the uncertain interval as the final trustworthiness of the proposition. If this value has the highest trustworthiness among the possible hypothesis, this assumption is the final decision result.

2.2. Posterior Probability Least Squares Support Vector Machine

In the study, the fault samples collected are limited, while support vector machine (SVM) and LSSVM can obtain high diagnosis accuracy based on small sample data. Moreover, the speed of LSSVM is faster than that of SVM, so LSSVM is selected to be the initial classifier to judge the state. As the input parameters of the D–S evidence fusion are basic probability assignments in all classification spaces, the hard output (whether or not) of the traditional classifier has to be converted to a soft one (probability) [33]; that is, the output of the classifier must be changed to the posterior probability output. For the two-class problem, the posterior probability can be calculated using the sigmoid function to map the output $f(x)$ $(+1, -1)$ of the LSSVM to the $[0, 1]$ interval. Assuming that the probability is consistent with the sigmoid distribution, the posterior probability can be calculated [34]:

$$p(y = 1/f) = \frac{1}{1 + exp(Af + B)} \qquad (6)$$

$$p(y = -1/f) = 1 - p(y = 1/f) \tag{7}$$

where f is the classification result of the standard LSSVM, $p(y = 1/f)$ is the probability when the classification is correct under the condition that the output value is f, $p(y = -1/f)$ is the probability when the classification is wrong under the condition that the output value is f, and A and B are parameters. So, the key to calculating the posterior probability is to obtain parameters A and B. The posterior probability least squares support vector machine model is usually established by first establishing a standard LSSVM model, and then obtaining A and B on the training set (f_i, t_i), where t_i is the target probability output of the standard LSSVM:

$$t_i = \begin{cases} \frac{N_+ + 1}{N_+ + 2}, f_i = +1 \\ \frac{1}{N_- + 2}, f_i = -1 \end{cases} \tag{8}$$

where N_+ is the number of positive samples; N_- is the number of negative samples; and the problem of obtaining parameters A and B is to solve the minimum likelihood optimization problem of the following, i.e.,

$$min\left\{-\sum_{i=1}^{n}[t_i \log(p_i) + (1 - t_i)\log(1 - p_i)]\right\} \tag{9}$$

where

$$p_i = \frac{1}{1 + exp(Af_i + B)} \tag{10}$$

The Hessian matrix for solving (9) is as follows:

$$H(z) = \begin{bmatrix} \sum_i f_i^2 p_i(1 - p_i) & \sum_i f_i p_i(1 - p_i) \\ \sum_i f_i p_i(1 - p_i) & \sum_i p_i(1 - p_i) \end{bmatrix} \tag{11}$$

In order to get the minimum value of (9), the Hessian matrix must be positively determined. So, A and B are finally obtained by solving all the eigenvalues of the matrix that are greater than zero. The posterior probability can be obtained.

It is proved that the posterior probability least squares support vector machine with sigmoid function works well in practical applications [35], but this method can only be used for the two-class problem. The main methods for extending LSSVM from two-class to multi-class are the "one-versus-one" and "one-versus-all" methods. The Platt algorithm calculates the probability formula for each classifier as follows, where p_m is the probability that sample x belongs to the i-th class [35]:

$$p_m = (y = m/x) = \frac{1}{1 + exp(A_m f(x) + B_m)} \tag{12}$$

2.3. The Improved Artificial Bee Colony

There are three kinds of kernel functions commonly used in LSSVM: linear kernel function, polynomial kernel function, and radial basis function (RBF) ($K(x_i, x_i) = exp(-\|x_i - x_i\|^2 / \sigma^2)$, where σ is the kernel width). Many studies and experiments [36] show that, compared with other kernel functions, RBF can map the original space into an infinite dimensional space and find the hyperplane better. It is a better choice as the kernel function. Therefore, it is necessary to select the regularization parameter γ (necessary for LSSVM, determining the trade-off between the training error minimization and smoothness) and the kernel squared bandwidth σ^2.

Choosing a better parameter value can greatly improve the performance of the LSSVM classifier and the accuracy of diagnosis. At present, the commonly used methods include trial and error, cross validation, grid search, and intelligent optimization algorithm [37].

Among them, the trial and error method not only consumes time and energy, but also the choice of parameters is greatly affected by subjective factors; the cross validation method divides the data set into training, validation, and testing, and different proportions will lead to different optimal models and optimal parameters; and the grid search method optimizes the model according to the set step size in the upper and lower limits of parameters, and then determines the optimal parameters, so the search speed is too slow and the precision is not high. Therefore, the advantages of the intelligent optimization algorithm are highlighted. It realizes the optimal distribution of food by simulating the behavior of animals in the population (interact information and cooperation among individuals). A swarm intelligence optimization algorithm is easy to implement and has high efficiency, so it is applied to the parameter optimization process of LSSVM.

Swarm intelligence optimization algorithms include genetic algorithm, particle swarm optimization, artificial fish swarm algorithm, artificial bee colony algorithm, and so on. Among them, artificial bee colony algorithm (ABC) is an optimization algorithm proposed in recent years, which not only has good optimization ability, but also controls less parameters in the process. Furthermore, it is simple, flexible, and easier to implement. The research [38] shows that the optimization performance of ABC is better than that of genetic algorithm and particle swarm algorithm, and the classification diagnosis accuracy of LSSVM optimized by ABC is higher than that of LSSVM optimized by genetic algorithm and particle swarm algorithm.

However, ABC has some shortcomings, such as slow convergence speed in the later stage of operation and the fact that it is easy to fall into local optimum. Therefore, in this paper, on the one hand, chaotic initialization is introduced in the artificial bee colony algorithm, which is used to initialize the population position to improve the diversity of the population and the ergodicity of the population search process. On the other hand, in the collecting bees stage of the artificial bee colony algorithm, the bees are divided into two parts: one part collects the optimal information of the region according to the original algorithm, and the other does Lévy flight around the global optimal solution to improve their global search capabilities. At the same time, in the observing bees stage, a search strategy based on the current local optimal solution (called pbest) is adopted to improve the local search ability of the algorithm.

(1) The logistic chaotic map is proposed to initialize the population. The equation for the logistic chaotic map is as follows:

$$y_{t+1} = \mu y_t (1 - y_t) \quad t = 0, 1, \cdots, l \qquad (13)$$

In the formula, $y_t \in (0,1)$, t is the number of iterations of the chaotic sequence, μ is the control parameter of the chaotic sequence, and the value range is $[3.75, 4]$ [39].

(2) Lévy flight was introduced in the evolution strategy to improve the performance of the algorithm and achieve good results [39]. The calculation method is based on

$$L(\alpha) = \left[\frac{\Gamma(1+\alpha)sin\left(\frac{\pi\alpha}{2}\right)}{\Gamma\left(\frac{1+\alpha}{2}\right)\alpha 2^{\left(\frac{\alpha-1}{2}\right)}} \right]^{1/\alpha} \qquad (14)$$

where α is the characteristic index, which usually satisfies $0 < \alpha < 2$. $\Gamma(\cdot)$ is the Gamma function defined as

$$\Gamma(z) = \int_0^\infty t^{z-1} e^{-t} dt.$$

Its update equation is as follows:

$$v_{ij} = x_{ij} + \alpha (x_{ij} - x_{best}) L(\alpha) \qquad (15)$$

where α is the step length, which usually meets the standard normal distribution, and $L(\cdot)$ is the random search path for Lévy flight.

(3) In the observing bees stage, for any current solution in each generation, the top $p\%$ solutions are randomly selected among all current solutions, and the best one (called pbest) can be used to balance global search capabilities and local development capabilities. The neighborhood search formula is as follows:

$$v_{ij} = x_{ij} + \varphi_{ij}\left(x_{ij} - x_{kj}\right) + \varnothing_{ij}(x_{randp,j}^{best} - x_{ij}) \tag{16}$$

where $k \in \{1, 2, \cdots, S_N\}$, S_N is the number of solutions for the bee colony, $j \in \{1, 2, \cdots D\}$, D is the dimension of the optimization problem, $k \neq i$, $\varphi_{ij} \in [-1, 1]$, and $\varnothing_{ij} \in [0, 1.5]$.

3. Specific Steps for Misalignment Diagnosis

D–S evidence theory is used to carry out the fault diagnosis of wind turbines. The specific steps are as follows.

(1) Identify the frame of discernment of the fault diagnosis system

The frame of discernment is the common faults of the wind turbines misalignment in the study. At the same time, the normal working state of the unit is added. So, the frame of discernment is expressed as follows: {normal, parallel misalignment, angular misalignment and integrated misalignment}.

(2) Determination of evidence

The posterior probability least squares support vector machines are trained by the vibration signal, the temperature signal, and the stator current signal feature vectors separately. The hard outputs of the traditional LSSVM are mapped to the [0, 1] interval using the sigmoid function. The soft outputs of the transformation are used as evidences for D–S evidence theory.

(3) Determination of basic probability assignment function, belief function, and plausibility function

The three least squares support vector machines give the probability vectors of all the classifications on the entire identification framework respectively, and the probability vectors to be directly used as the basic probability assignments, belief function, and plausibility function can be obtained by calculation.

(4) Evidence synthesis and diagnosis

According to Dempster's law, the probability vectors directly participate in the evidence fusion process. After the final probability vector is given, the final diagnosis result based on the probability vector after fusion can be obtained.

Figure 2 summarizes process of D–S evidence-based misalignment diagnosis.

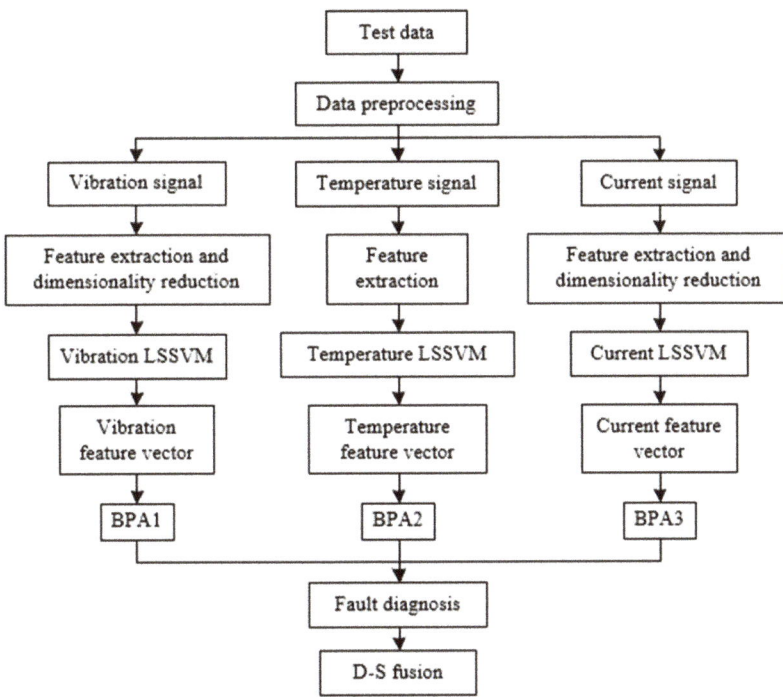

Figure 2. Process of Dempster–Shafer (D–S) evidence fusion. LSSVM, least square support vector machine; BPA, basic probability assignment.

4. The Simulation Case Studies of Misalignment Fault Diagnosis

The simulation wind turbine system is established by ADAMS 2013, MATLAB R2014a, and Ansys 17.0. The three-dimensional (3D) model of the 1.5 MW wind turbine is established using SolidWorks, and then it is imported into ADAMS 2013, where the Marker point is moved according to the type and degree of misalignment; that is, parallel misalignment is simulated by making the center of mass deviate from the center of rotation for a certain distance; angle misalignment is simulated by rotating the marker a certain angle around the y-axis, and placing the rotation axis of the coupling relative to the ground on the z-axis of the Marker point; and integrated misalignment is simulated by adding the parallel misalignment and angle misalignment in the local coordinate system (maker) of the left half coupling at the same time. The correctness of the models has been verified in the literature [40]. The vibration signals were extracted under the input speeds of 81.3°/s, using step function as the input of ADAMS, the simulation time is 1.5 s, and simulation steps are 6000 steps. The wind turbine models and its control system are established by SIMULINK/MATLAB, where the stator current was sampled at the same speed at which the vibration signal was sampled, and the sample frequency is 200 kHz. The correctness of the models has been verified in the literature [41]. After that, the high-speed gear shaft and the main shaft of the generator are introduced into HyperMesh to divide the grid. Then, the model is imported into Ansys Workbench to get the corresponding temperature signals (details in the literature [42]). In this paper, 100 samples are taken for each of the four types of diagnostic states (normal, parallel misalignment, angular misalignment, and integrated misalignment), of which 60 are for training and 40 are for testing. So, there are 240 (60 × 4) samples in the training set and there are 160 (40 × 4) samples in the testing set.

4.1. Data Processing

After the vibration signal, temperature signal, and stator current signal under four working conditions are collected, in order to make better use of them and get good diagnosis results, the feature indexes in the time, frequency, and time–frequency domain are extracted. Table 2 shows a 21-dimension mixed feature library of the vibration signal.

Table 2. Mixed feature library of vibration signals. IEMD, image extended empirical mode decomposition.

Feature Library	Feature	Index
Mixed-domain feature library	Time domain	Root mean square, square root amplitude, variance, standard deviation, kurtosis, waveform index, peak index, pulse index, margin index, kurtosis index
	Frequency domain	Center of gravity frequency, mean square frequency, frequency variance
	Time–frequency domain	The first eight energy entropy of the IMF (intrinsic mode function) component of IEMD decomposition

Suppose signal x ($x_0, x_1, x_2, \cdots, x_{N-1}$) is a discrete time series with a finite length, the calculation formulas of time domain characteristic indexes are shown in Table 3, where $\bar{x}$ is the mean value of the signal, x' is the average amplitude, and x_p is the peak value of the signal.

Table 3. Time domain characteristic index.

	Time Domain Index	Calculation Formula
Dimensional indicators	Root mean square	$x_{rms} = (\frac{1}{N} \sum_{i=0}^{N-1} x_i^2)^{1/2}$
	Square root amplitude	$x_r = (\frac{1}{N} \sum_{i=0}^{N-1} |x_i|^{\frac{1}{2}})^2$
	Variance	$\delta = \frac{1}{N} \sum_{i=0}^{N-1} (x_i - \bar{x})^2$
	Standard deviation	$x_{std} = (\frac{1}{N} \sum_{i=1}^{N} (x_i - \bar{x})^2)^{1/2}$
	Kurtosis	$\beta = \frac{1}{N} \sum_{i=1}^{N} x_i^4$
Dimensionless index	Waveform index	$K = x_{rms}/x'$ $(x' = \frac{1}{N} \sum_{i=0}^{N-1} |x_i|)$
	Peak index	$C = x_p/x_{rms}$
	Pulse index	$I = x_p/x'$
	Margin index	$L = x_p/x_r$
	Kurtosis index	$K_r = \frac{\frac{1}{N} \sum_{i=1}^{N} x_i^4}{x_{rms}^4}$

In signal analysis, power spectrum analysis is usually used to extract the frequency domain index. Center of gravity frequency, mean square frequency, root mean square frequency, and frequency variance are commonly used. The sampling frequency is set as f_s, and the calculation formula of each index is shown in Table 4, where $S(\omega)$ is the power spectrum of discrete time series, $S(\omega) = X(\omega) \cdot \overline{X(\omega)}$, $X(\omega) = \sum_{i=0}^{N-1} x(i) e^{-j\pi\omega}$, ω is the angular frequency.

Table 4. Frequency domain characteristic index.

Frequency Domain Index	Calculation Formula
Center of gravity frequency	$FC = \frac{1}{2\pi f_s} \cdot \frac{\int_0^\pi \omega S(\omega)d\omega}{\int_0^\pi S(\omega)d\omega}$
Mean square frequency	$MSF = \frac{1}{4\pi^2 f_s^2} \cdot \frac{\int_0^\pi \omega^2 S(\omega)d\omega}{\int_0^\pi S(\omega)d\omega}$
Root mean square frequency	$RMSF = (\frac{1}{4\pi^2 f_s^2} \cdot \frac{\int_0^\pi \omega^2 S(\omega)d\omega}{\int_0^\pi S(\omega)d\omega})^{\frac{1}{2}}$
Frequency variance	$VF = \frac{1}{4\pi^2 f_s^2} \cdot \frac{\int_0^\pi (\omega - 2\pi f_s FC) S(\omega)d\omega}{\int_0^\pi S(\omega)d\omega}$

Time–frequency analysis is a fault diagnosis method that combines the law and reason of frequency changing with time. In this paper, image extended empirical mode decomposition (IEMD) is used to process the vibration signal, and dual tree complex wavelet transform (DTCWT) is used to process the stator current signal (see the literature [43] for details).

The gearbox tooth temperature T_1 and the generator rotor shaft temperature T_2 are selected as the characteristic values of the temperature signal. Construct a two-dimensional vector of the temperature signal: $X = [T_1, T_2]$.

Table 5 is a mixed feature library with a total of 29 dimensions in the time domain, frequency domain, and time–frequency domain of the stator current signal (see the literature [41] for details).

Table 5. Mixed feature library of stator current signals.

Feature Library	Feature	Index
Mixed-domain feature library	Time domain	Root mean square, square root amplitude, variance, standard deviation, kurtosis, waveform index, peak index, pulse index, margin index, kurtosis index
	Frequency domain	Center of gravity frequency, mean square frequency, root mean square frequency, frequency variance
	Time–frequency domain	Sample entropy 1–5, energy entropy H_1, H_2, H_3, H_4, H_5, spectral kurtosis a_1, a_2, a_3, a_4, a_5

In order to eliminate the influence of different input dataset dimensions and large numerical differences, the original dataset is normalized, i.e.,

$$y = \frac{y_{max} - y_{min}}{x_{max} - x_{min}} * (x - x_{min}) + y_{min} \qquad (17)$$

where x is the value to be normalized, y_{min} is the lower bound of the normalized interval, and y_{max} is the upper bound of the normalized interval. In this paper, $y_{min} = 0$, $y_{max} = 1$, and the vector is normalized by column.

Because of the high dimensionality of the constructed vectors of the vibration signal and the stator current signal, not only does the amount of calculation increase, but also some difficulties are brought to fault diagnosis [44]. In order to make better use of various information and obtain good diagnostic results, the feature vectors are subjected to dimensionality reduction using t-SNE.

t-SNE based on conditional probability retains the similarity between high-dimensional and low dimensional space data and adopts symmetric objective function, and t distribution in low-dimensional space replaces Gaussian distribution, which solves the problem of crowding and clear visualization in low-dimensional space [45]. Its implementation steps are as follows:

(1) Define a high-dimensional data set: $x = \{x_1, x_2, \cdots, x_n\}$.
(2) Compute the complexity parameter of the value equation c:

$$c = \sum_i \sum_j p_{ij} \log \frac{p_{ij}}{q_{ij}} \tag{18}$$

$$perp(p_i) = 2^{H(p_i)} \tag{19}$$

$$H(p_i) = -\sum_j p_{j/i} \log_2^{p_{j/i}} \tag{20}$$

where p_i is the conditional probability of data points (other than x_i) with respect to x_i, $p_{j/i}$ is the conditional probability of high-dimensional data, p_{ij} is the joint probability density in the high-dimensional space, and q_{ij} is the joint probability density in the low-dimensional mapping space.

(3) Define the optimization parameters: the number of iterations T, the learning rate η, and the momentum factor at the tth (t ≤ T) iteration $\alpha(t)$ (0 < $\alpha(t)$ < 1). The value equation c is learned by the gradient descent method, and the low-dimensional mapping of the high-dimensional data is finally obtained:

$$\frac{\delta c}{\delta y_i} = 4 \sum_j (p_{ij} - q_{ij})(y_i - y_j)(1 + \parallel y_i - y_j \parallel^2)^{-1} \tag{21}$$

where y_i and y_j are the mapping of the high-dimensional data x_i and x_j in the low-dimensional space.

In order to speed up the optimization process and prevent trapping into local minima, a relatively large momentum condition is imposed on the descent process. The current gradient value is summed to the previous gradient value for each iteration and then decays exponentially to determine the coordinates of the low-dimensional data. The momentum formula is as follows:

$$y^{(t)} = y^{(t-1)} + \eta \frac{\delta c}{\delta y} + \alpha(t)\left(y^{(t-1)} - y^{(t-2)}\right) \tag{22}$$

where y is the data in the low-dimensional space.

4.2. The Fault Diagnosis Results

In this paper, "one-versus-all" is used to extend LSSVM from two classifications to multiple classifications. That is, each time, one fault is selected as one type, and the rest of the states are selected as another type. In order to produce the posterior probabilities of the four classifications in the vibration feature space, four two-class LSSVM are constructed, and each LSSVM calculates a set of A and B, and then the corresponding posterior probability is calculated according to (5) and (6). In the same way, the probability vectors of the temperature and stator current signal classifiers for the four states can be obtained as the BPA of D–S evidence fusion.

The five-dimensional feature vectors of the vibration signal after t-SNE dimensionality reduction are used as the inputs, and the four working conditions of the transmission system are used as outputs to train the LSSVM, which is optimized by the improved artificial bee colony algorithm. The parameters of the four two-classification LSSVM in the vibration feature space are shown in Table 6. Four samples are selected, such as samples 5, 44, 82, and 130, and the corresponding BPA1 calculated is shown in Table 7.

Table 6. Parameters in four vibration least square support vector machine (LSSVM).

LSSVM	Γ	σ^2	A	B
(normal, the rest)	84.2784	31.1601	−10.8211	−4.2378
(parallel misalignment, the rest)	85.8947	30	−7.4866	−1.9740
(angular misalignment, the rest)	36.0326	39.9616	−6.1128	−1.5637
(integrated misalignment, the rest)	99.4093	96.6015	−6.9786	−2.4798

Table 7. Basic probability assignment 1 (BPA1) of vibration LSSVM.

Sample Number	Normal	Parallel Misalignment	Angular Misalignment	Integrated Misalignment
5	0.8433	0.0314	0.0538	0.0715
44	0.0080	0.9246	0.0295	0.0379
82	0.0025	0.0070	0.8873	0.1032
130	0.0208	0.0147	0.0495	0.9150

The two-dimensional feature vectors of the temperature signal are used as the inputs, and the four operating states of the transmission system are used as the outputs to train the optimized LSSVM. The parameters of the four binary LSSVM in the temperature feature space are shown in Table 8. The BPA2 calculated from the same four samples is shown in Table 9.

Table 8. Parameters in four temperature LSSVM.

LSSVM	Γ	σ^2	A	B
(normal, the rest)	97.4952	89.7017	−3.0178	−0.4262
(parallel misalignment, the rest)	98.3829	30	−2.8893	0.1963
(angular misalignment, the rest)	46.3907	88.9951	−3.9974	−0.3622
(integrated misalignment, the rest)	93.6931	96.3611	−2.4749	0.3805

Table 9. BPA2 of temperature LSSVM.

Sample Number	Normal	Parallel Misalignment	Angular Misalignment	Integrated Misalignment
5	0.7418	0.0387	0.0228	0.1967
44	0.1314	0.7670	0.0243	0.0773
82	0.3644	0.0357	0.5549	0.0450
130	0.4365	0.0394	0.0220	0.5021

The four-dimensional vectors after the dimensionality reduction of the stator current signal are used as inputs, and the four operating states of the transmission system are as outputs to train the optimized LSSVM. The parameters of the four two-class LSSVM in the stator current feature space are shown in Table 10, and the BPA3 calculated by the same four samples is shown in Table 11.

Table 10. Parameters in four current LSSVM.

LSSVM	Γ	σ^2	A	B
(normal, the rest)	35.2345	53.0811	−4.858	−0.6011
(parallel misalignment, the rest)	87.5286	30	−3.2022	−0.1165
(angular misalignment, the rest)	100	30	−3.2755	−0.2514
(integrated misalignment, the rest)	77.1503	33.2639	−3.2803	−0.2146

Table 11. BPA3 of current LSSVM.

Sample Number	Normal	Parallel Misalignment	Angular Misalignment	Integrated Misalignment
5	0.8635	0.0440	0.0468	0.0457
44	0.0164	0.6193	0.3159	0.0484
82	0.0170	0.1236	0.8093	0.0501
130	0.0127	0.1093	0.0521	0.8259

Then, the probability assignments are calculated after the fusion of the three BPAs. The category with the highest degree of belief is selected as belonging to the class of the fusion model. Table 12 shows the basic and the fusion probability of the three LSSVM outputs for the selected test samples. Table 13 shows the fusion and classification results of the four test samples. Figure 3 shows the test samples' diagnosis results, in which "0" indicates normal operation, "1" indicates parallel misalignment, "2" indicates angular misalignment, and "3" indicates integrated misalignment.

Table 12. Probability assignment of three LSSVMs and fusion.

BPA 1	BPA 2	BPA3	D–S Evidence Fusion
[0.8433, 0.0314, 0.0538, 0.0715]	[0.7418, 0.0387, 0.0228, 0.1967]	[0.8635, 0.0440, 0.0468, 0.0457]	[0.9986, 0.0001, 0.0001, 0.0012]
[0.0080, 0.9246, 0.0295, 0.0379]	[0.1314, 0.7670, 0.0243, 0.0773]	[0.0164, 0.6193, 0.3159, 0.0484]	[0.0001, 0.9991, 0.0005, 0.0003]
[0.0025, 0.0070, 0.8873, 0.1032]	[0.3644, 0.0357, 0.5549, 0.0450]	[0.0170, 0.1236, 0.8093, 0.0501]	[0.0001, 0.0001, 0.9993, 0.0005]
[0.0208, 0.0147, 0.0495, 0.9150]	[0.4365, 0.0394, 0.0220, 0.5021]	[0.0127, 0.1093, 0.0521, 0.8259]	[0.0003, 0.0002, 0.0001, 0.9994]

Table 13. Fusion and classification results of four test samples.

D–S Evidence Fusion	Category	Is the Classification Correct?
[0.9986, 0.0001, 0.0001, 0.0012]	Normal	Yes
[0.0001, 0.9991, 0.0005, 0.0003]	Parallel misalignment	Yes
[0.0001, 0.0001, 0.9993, 0.0005]	Angular misalignment	Yes
[0.0003, 0.0002, 0.0001, 0.9994]	Integrated misalignment	Yes

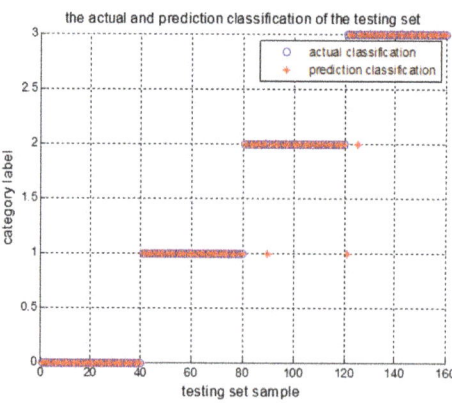

Figure 3. The diagnosis results of the testing set.

In order to better evaluate the performance of the fault diagnosis method, three indexes are adopted: the training set classification accuracy, the testing set classification accuracy, and the fault false alarm rate. The fault false alarm rate means that the fault does not actually occur, but the fault detection alarm is given by the detection system. The false alarm rate equals the number of false alarm samples divided by the total number of actual fault-free samples. Table 14 compares the results of the sample sets diagnosed by the indexes of a single signal (vibration, temperature, or current signal) with the D–S evidence fusion.

Table 14. Comparison of diagnostic results.

Signal Selection	Training Set Classification Accuracy	Testing Set Classification Accuracy	False Alarm Rate
Vibration signal	100% (240/240)	85.625% (137/160)	5% (2/40)
Temperature signal	90.8333% (218/240)	81.25% (130/160)	35% (14/40)
Current signal	99.5833% (239/240)	84.375% (135/160)	10% (4/40)
D–S evidence fusion	100% (240/240)	98.125% (157/160)	0% (0/40)

From Table 14, it can be seen that the accuracy of D–S fusion is higher than that of any single signal, and the failure false alarm rate is equal to zero, lower than others, which proves the advantage of information fusion in the diagnosis of wind turbine misalignment fault.

5. Experimental Verification of Platform

In this paper, the 1.5 kW misalignment experimental platform is used for experimental verification. The platform is shown in Figure 4a. It includes a generator, coupling, gearbox, driving motor, and so on. The speed of the driving motor is changed by a planetary gear reducer with a transmission ratio of 1:50 to simulate the wind blowing blade speed, then it is accelerated by a planetary gear with a transmission ratio of 40:1 and a spur gear with a transmission ratio of 1.5:1 to drive the generator. The generator can be adjusted by the support to create parallel or angular misalignment.

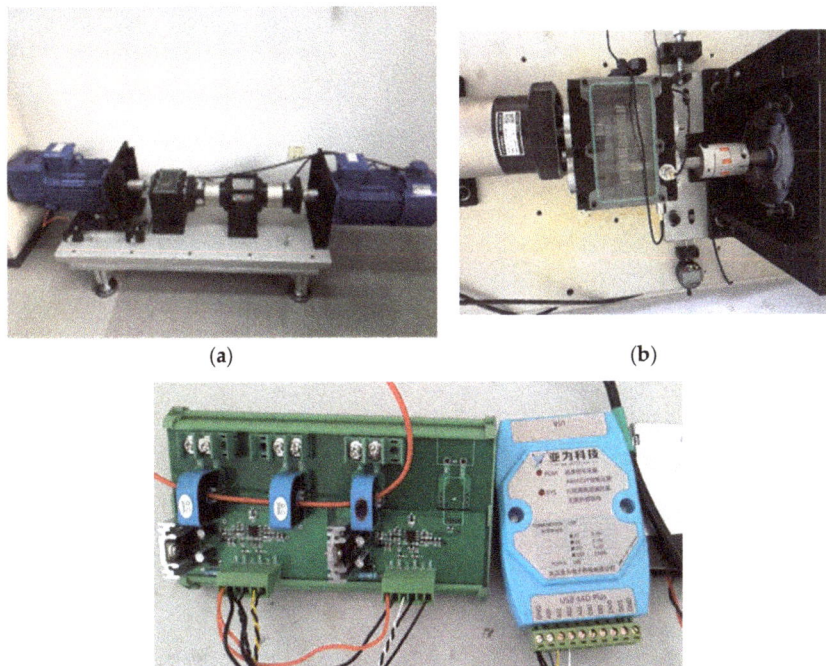

Figure 4. Experiment equipment. (**a**) The platform of wind turbine; (**b**) layout of vibration sensor; (**c**) current signal acquisition card USB 4AD Plus.

The vibration signal of the gearbox is obtained using the DFT5100 dynamic data collector from the acceleration sensor (ICP type) on the experimental platform (Figure 4b). The current signal is transmitted to the USB signal acquisition and recording platform through the signal acquisition card USB 4AD Plus (Figure 4c). In this paper, the rotation speed of the motor is set to 600 rpm; the sampling time is 10 s; and the sampling frequency of vibration and current is 1 kHz and 2 kHz, respectively. In the experiments, the temperature signal is easily affected by the operation time of the unit and the ambient temperature, and it cannot reflect the actual operating temperature of the wind turbine. Therefore, when fusing different signals by D–S evidence theory, we set the temperature signal to 0, regardless of its influence. Four groups for each working condition, with a total of 16 groups, are sampled on the platform. Some characteristic indexes of vibration and current signal are shown in Tables 15 and 16. The actual classification and diagnosis results of fusion signals and individual signals are shown in Figure 5. Table 17 is the calculation of two examples.

Table 15. Part of the characteristic index of the vibration signal.

Fault Type	Root Mean Square Value	Center of Gravity Frequency	IMF1 Energy Entropy	IMF2 Energy Entropy
Normal	0.0286	−118.3859	0.3671	0.0991
	0.0270	−184.0340	0.3461	0.1265
	0.0288	−308.9050	0.3678	0.1624
	0.0626	−993.2476	0.3524	0.3675
Parallel misalignment	0.0248	−272.4819	0.3678	0.1229
	0.0258	−196.5053	0.3677	0.1201
	0.0253	−286.5526	0.3668	0.1769
	0.0607	−1082.4788	0.3678	0.3455
Angular misalignment	0.0266	−166.9377	0.3488	0.1005
	0.0296	−145.9158	0.3658	0.1083
	0.0280	−232.8465	0.3620	0.1347
	0.0569	−1052.415	0.3583	0.3677
Integrated misalignment	0.0284	−261.2838	0.3544	0.1615
	0.0342	−334.0774	0.3621	0.2021
	0.0311	−388.9565	0.3675	0.2138
	0.0670	−1138.6520	0.3603	0.3670

Table 16. Partial characteristic index of current signal.

Fault Type	Root Mean Square Value	Center of Gravity Frequency	Energy Entropy1	Sample Entropy1
Normal	2.4944	−0.0571	0.0001	0.6859
	2.4952	−0.0741	0.0003	0.8285
	3.5641	−0.1023	0.0008	0.9624
Parallel misalignment	2.4948	−0.1238	0.0001	0.6099
	2.5293	−0.2608	0.0004	0.7046
	2.7607	−0.4788	0.0015	0.9455
Angular misalignment	2.4990	−0.1062	0.0002	1.0642
	2.6908	−0.2794	0.0008	1.1659
	3.0569	−0.4415	0.0016	1.3677
Integrated misalignment	2.5051	−0.0524	0.0003	1.2666
	2.8986	−0.3861	0.0009	1.6857
	3.2670	−0.6520	0.0023	1.7670

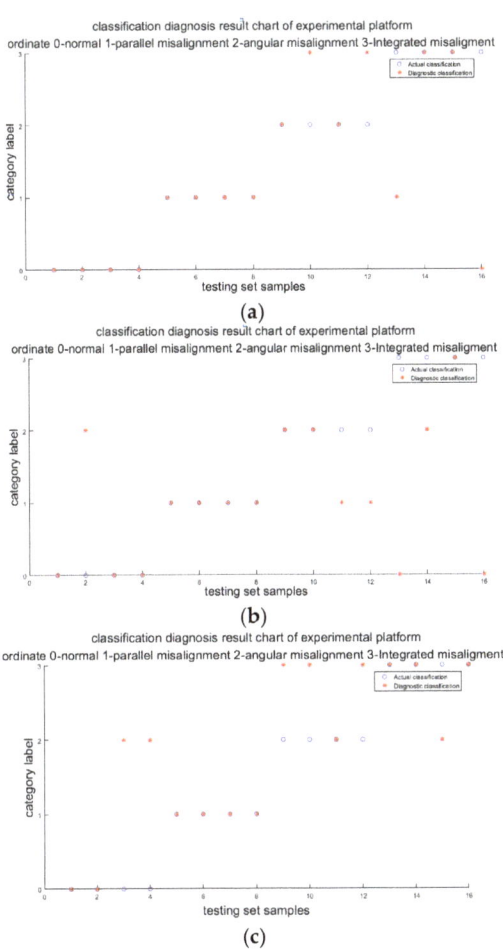

Figure 5. Diagnostic result. (**a**) Vibration signal + current signal; (**b**) vibration signal; (**c**) current signal.

Table 17. Basic probability assignment of two kinds signals and probability after fusion.

Probability Value	Normal	Parallel Misalignment	Angular Misalignment	Integrated Misalignment
Vibration signal	0.5161	0.1663	0.1510	0.1666
	0.1770	0.1927	0.4194	0.2109
Current signal	0.8209	0.0737	0.0181	0.0873
	0.7705	0.0566	0.1063	0.0666
Fusion signal	0.9348	0.0281	0.0050	0.0321
	0.6623	0.0530	0.2164	0.0683

It can be seen from Figure 5 that the classification accuracy of the testing set is 75%, while that of the single vibration signal is 62.5% and that of the single current signal is 62.5%, which indicates that the accuracy of the diagnosis is improved by using the D–S decision fusion method with multi-source signals as the diagnosis information. In addition, the reason the classification accuracy of the experimental results is much lower than that of the simulation results is that there is no temperature signal in the D–S evidence theory fusion. It can be seen from Table 17 that the first sample is correctly identified using either the single signal or fusion signal, while the second sample is mistakenly diagnosed as angle misalignment using only the vibration signal, but is correctly identified by D–S fusion.

6. Conclusions

This paper proposes an integrated fault diagnosis method for wind turbine transmission system misalignment based on information decision fusion. The method uses multiple sources of signal including vibration signal, temperature signal, and stator current signal as the original source, and extracts different features from their time domain, frequency domain, and time–frequency domain. t-SNE is used to eliminate the correlation of characteristic values of the vibration signal and the stator current signal. Three posterior probability least squares support vector machines optimized using improved artificial bee colony algorithm are constructed respectively. The output probabilities of least squares support vector machines are used as the basic probability distribution of evidence fusion, and the fault diagnosis is completed by D–S synthesis and decision rules. Finally, the simulation experiments and platform verification show that the D–S evidence fusion model has higher diagnostic accuracy than the non-fusion model for the wind turbine misalignment fault.

Author Contributions: Conceptualization, Y.X. and Y.W.; software, Y.W.; validation, M.L.; writing—original draft preparation, Y.W. and Y.X.; writing—review and editing, L.Z. and J.X.; project administration, Y.X. All authors have read and agreed to the published version of the manuscript.

Funding: This research was funded by the National Natural Science Foundation of China (51577008).

Conflicts of Interest: The authors declare no conflict of interest.

References

1. Junior, P.R.; Fischetti, E.; Araújo, V.G.; Peruchi, R.S.; Aquila, G.; Rocha, L.C.S.; Lacerda, L.S. Wind power economic feasibility under uncertainty and the application of ANN in sensitivity analysis. *Energies* **2019**, *12*, 2281. [CrossRef]
2. Liu, Z.; Zhang, L.; Carrasco, J. Vibration analysis for large-scale wind turbine blade bearing fault detection with an empirical wavelet thresholding method. *Renew. Energy* **2020**, *146*, 99–110. [CrossRef]
3. Hu, A.; Yan, X.; Xiang, L. A new wind turbine fault diagnosis method based on ensemble intrinsic time-scale decomposition and WPT-fractal dimension. *Renew. Energy* **2015**, *83*, 767–778. [CrossRef]
4. Zheng, H. Case analysis of wind turbine vibration monitoring. *Wind Energy* **2014**, *2017*, 88–92. (In Chinese) [CrossRef]
5. Liao, M.; Liang, Y.; Wang, S.; Wang, Y. Analysis of misalignment of wind turbines. *Mech. Sci. Technol. Aerosp. Eng.* **2011**, *2011*, 173–180. (In Chinese with English abstract) [CrossRef]
6. Liu, Z.; Zhang, L. A review of failure modes, condition monitoring and fault diagnosis methods for large-scale wind turbine bearings. *Measurement* **2020**, *149*, 107002. [CrossRef]
7. Simani, S.; Fantuzzi, C. Dynamic system identification and model-based fault diagnosis of an industrial gas turbine prototype. *Mechatronics* **2006**, *16*, 341–363. [CrossRef]
8. Tang, H.; Lam, K.-M.; Shum, K.-M.; Li, Y. Wake effect of a horizontal axis wind turbine on the performance of a downstream turbine. *Energies* **2019**, *12*, 2395. [CrossRef]
9. Zhao, Q.; Han, T.; Jiang, D.; Yin, K. Application of variational mode decomposition to feature isolation and diagnosis in a wind turbine. *J. Vib. Eng. Technol.* **2019**, *7*, 639–646. [CrossRef]
10. Abdalla, G.M.; Tian, X.; Zhen, D.; Gu, F.; Ball, A.; Chen, Z. Misalignment diagnosis of a planetary gearbox based on vibration analysis. In Proceedings of the 21st International Congress on Sound and Vibration, Beijing, China, 13–17 July 2014; Volume 4, pp. 2775–2783. Available online: https://www.researchgate.net/publication/286770020 (accessed on 20 January 2021).
11. Huang, Q.; Jiang, D.X.; Hong, L.Y. Application of Hilbert-Huang transform method on fault diagnosis for wind turbine rotor. *Key Eng. Mater.* **2009**, *413*, 159–166. [CrossRef]
12. An, M.-S.; Kang, D.-S. Application of modified empirical mode decomposition method to fault diagnosis of offshore wind turbines. *Int. J. Multimed. Ubiquitous Eng.* **2016**, *11*, 67–80. [CrossRef]
13. Villa, L.F.; Reñones, A.; Perán, J.R.; De Miguel, L.J. Statistical fault diagnosis based on vibration analysis for gear test-bench under non-stationary conditions of speed and load. *Mech. Syst. Signal Process.* **2012**, *29*, 436–446. [CrossRef]
14. He, G.; Ding, K.; Li, L.; Deng, R. Vibration test and analysis of transmission chain of wind turbine based on double-elastic support. *J. South China Univ. Technol. (Nat. Sci. Ed.)* **2014**, *2014*, 90–97. (In Chinese with English abstract) [CrossRef]
15. Bossé, E.; Rogova, G.L. *Information Quality in Information Fusion and Decision Making*; Information Fusion and Data Science; Springer: Berlin/Heidelberg, Germany, 2019; pp. 1–49.

16. Mönks, U. *Information Fusion under Consideration of Conflicting Input Signals*; Technologien für die intelligente Automation; Springer: Berlin/Heidelberg, Germany, 2017; pp. 11–35.
17. Wald, L. Some new terms of reference in data fusion. *IEEE Trans. Geosci. Remote Sens.* **1999**, *37*, 1190–1193. [CrossRef]
18. Zeng, Q.; Liu, X. Research on Bayesian classification algorithm. *Biotechnol. World* **2015**, *35*, 253–255. (In Chinese)
19. Dubois, D.; Liu, W.; Ma, J.; Prade, H. The basic principles of uncertain information fusion. An organised review of merging rules in different representation frameworks. *Inf. Fusion* **2016**, *32*, 12–39. [CrossRef]
20. Yao, P. *Research on Fuzzy Multiple Attribute Decision Making Method for Process Industry Fault Diagnosis*; Qilu University of Technology: Jinan, China, 2019. (In Chinese with English abstract)
21. Li, W.-W.; Wu, C. A multicriteria interval-valued intuitionistic fuzzy set topsis decision-making approach based on the improved score function. *J. Intell. Syst.* **2016**, *25*, 239–250. [CrossRef]
22. Su, C.-H.; Chen, T.-L.; Cheng, C.-H.; Chen, Y.-C. Forecasting the stock market with linguistic rules generated from the minimize entropy principle and the cumulative probability distribution approaches. *Entropy* **2010**, *12*, 2397–2417. [CrossRef]
23. Wang, M.; Zhang, Z.; Pei, P. Application of information fusion technology in fault diagnosis of large generators. *Comput. Simul.* **2012**, *29*, 349–352. (In Chinese with English abstract) [CrossRef]
24. Li, Y.; Xu, Y.; Chen, G.; Miao, R.; Yu, J. Improvement and application of D-S evidence theory in multi-sensor fault diagnosis. *J. Southeast Univ. (Nat. Sci. Ed.)* **2011**, *41*, 102–106. (In Chinese with English abstract) [CrossRef]
25. Jiang, W.; Wu, S. Multi-data fusion fault diagnosis method based on SVM and evidence theory. *Chin. J. Sci. Instrum.* **2010**, *31*, 1738–1743. (In Chinese with English abstract) [CrossRef]
26. Hu, J.; Yu, Z.; Zhai, X.; Peng, J.; Ren, L. Research on fault diagnosis and fusion diagnosis of aero-engine rotor based on improved D-S evidence theory. *Acta Aeronaut. Astronaut. Sin.* **2014**, *35*, 436–443. (In Chinese with English abstract)
27. Tian, Y.; Liu, S.; Jing, Y.; Yang, Y. Fault prediction of wind turbines based on D-S evidence fusion. *Comput. Mod.* **2017**, *2017*, 57–61. (In Chinese with English abstract) [CrossRef]
28. Kang, J.; Gu, Y.; Li, Y. Multi-sensor information fusion algorithm based on DS evidence theory. *J. Chin. Inert. Technol.* **2012**, *20*, 670–673. (In Chinese with English abstract) [CrossRef]
29. Zhou, H.; Li, S. The combination of support vector machine and evidence theory in information fusion. *J. Transduct. Technol.* **2008**, *21*, 1566–1570. (In Chinese with English abstract) [CrossRef]
30. Boudraa, A.O.; Bentabet, L.; Salzenstein, F.; Guillon, L. Dempster-Shafer's basic probability assignment based on fuzzy membership functions. *ELCVIA* **2004**, *4*, 1–10. Available online: https://www.researchgate.net/publication/39087167 (accessed on 20 February 2021). [CrossRef]
31. Haenni, R. Shedding new light on Zadeh's criticism of Dempster's rule of combination. In Proceedings of the 2005 7th International Conference on Information Fusion, Philadelphia, PA, USA, 25–28 July 2005; Volume 2, p. 33.
32. Li, W.; Guo, K. Combination rules of D-S evidence theory and conflict problem. *Syst. Eng. Theory Pract.* **2010**, *30*, 1422–1432. (In Chinese with English abstract)
33. Zhang, X.; Xiao, X.; Xu, G. Weighted posterior probability output for support vector machines. *J. Tsinghua Univ. (Sci. Tech.)* **2007**, *47*, 1689–1691. (In Chinese with English abstract) [CrossRef]
34. Wang, J.P.; De Lin, S.; Bao, Z.F. Neural network and D-S evidence theory based condition monitoring and fault diagnosis of drilling. *Appl. Mech. Mater.* **2012**, *249*, 481–486. [CrossRef]
35. Lin, H.-T.; Lin, C.-J.; Weng, R.C. A note on Platt's probabilistic outputs for support vector machines. *Mach. Learn.* **2007**, *68*, 267–276. [CrossRef]
36. Zhao, F. Detection method of LSSVM network intrusion based on hybrid kernel function. *Mod. Electron. Technol.* **2015**, *38*, 97–99. (In Chinese with English abstract)
37. Yuan, X.; Wang, P.; Yuan, Y.; Huang, Y.; Zhang, X. A new quantum inspired chaotic artificial bee colony algorithm for optimal power flow problem. *Energy Convers. Manag.* **2015**, *100*, 1–9. [CrossRef]
38. Yi, Y.; He, R. A novel artificial bee colony algorithm. In Proceedings of the 2014 Sixth International Conference on Intelligent Human-Machine Systems and Cybernetics, Hangzhou, China, 26–27 August 2014; pp. 271–274.
39. Chen, W.; Xiao, Y. An improved ABC algorithm and its application in bearing fault diagnosis with EEMD. *Algorithms* **2019**, *12*, 72. [CrossRef]
40. Xiao, Y.; Kang, N.; Hong, Y.; Zhang, G. Misalignment fault diagnosis of DFWT based on IEMD energy entropy and PSO-SVM. *Entropy* **2017**, *19*, 6. [CrossRef]
41. Xiao, Y.; Hong, Y.; Chen, X.; Chen, W. The application of dual-tree complex wavelet transform (DTCWT) energy entropy in misalignment fault diagnosis of doubly-fed wind turbine (DFWT). *Entropy* **2017**, *19*, 587. [CrossRef]
42. Zhang, G. Thermal Characteristics Analysis of High Speed Transmission System of Wind Turbines. Master's Thesis, Beijing Jiaotong University, Beijing, China, 2017. (In Chinese with English abstract)
43. Xiao, Y.; Wang, Y.; Ding, Z. The Application of heterogeneous information fusion in misalignment fault diagnosis of wind turbines. *Energies* **2018**, *11*, 1655. [CrossRef]
44. Shao, R.; Hu, W.; Wang, Y.; Qi, X. The fault feature extraction and classification of gear using principal component analysis and kernel principal component analysis based on the wavelet packet transform. *Measurement* **2014**, *54*, 118–132. [CrossRef]
45. Zhang, Y.; Li, Y. Fisher information metric based on stochastic neighbor embedding. *J. Beijing Univ. Technol.* **2016**, *42*, 863–869. (In Chinese with English abstract)

Article

A New Deep Dual Temporal Domain Adaptation Method for Online Detection of Bearings Early Fault

Wentao Mao [1,2,*,†], Bin Sun [1,†] and Liyun Wang [1]

1. School of Information Engineering, Zhengzhou University of Industrial Technology, Zhengzhou 451100, China; 121114@htu.edu.cn (B.S.); zzwly0428@163.com (L.W.)
2. School of Computer and Information Engineering, Henan Normal University, Xinxiang 453007, China
* Correspondence: maowt@htu.edu.cn; Tel.: +86-150-3730-1821
† These authors contributed equally to this work.

Abstract: With the quick development of sensor technology in recent years, online detection of early fault without system halt has received much attention in the field of bearing prognostics and health management. While lacking representative samples of the online data, one can try to adapt the previously-learned detection rule to the online detection task instead of training a new rule merely using online data. As one may come across a change of the data distribution between offline and online working conditions, it is challenging to utilize the data from different working conditions to improve detection accuracy and robustness. To solve this problem, a new online detection method of bearing early fault is proposed in this paper based on deep transfer learning. The proposed method contains an offline stage and an online stage. In the offline stage, a new state assessment method is proposed to determine the period of the normal state and the degradation state for whole-life degradation sequences. Moreover, a new deep dual temporal domain adaptation (DTDA) model is proposed. By adopting a dual adaptation strategy on the time convolutional network and domain adversarial neural network, the DTDA model can effectively extract domain-invariant temporal feature representation. In the online stage, each sequentially-arrived data batch is directly fed into the trained DTDA model to recognize whether an early fault occurs. Furthermore, a health indicator of target bearing is also built based on the DTDA features to intuitively evaluate the detection results. Experiments are conducted on the IEEE Prognostics and Health Management (PHM) Challenge 2012 bearing dataset. The results show that, compared with nine state-of-the-art fault detection and diagnosis methods, the proposed method can get an earlier detection location and lower false alarm rate.

Keywords: fault detection; deep learning; transfer learning; anomaly detection; bearing

Citation: Mao, W.; Sun, B.; Wang, L. A New Deep Dual Temporal Domain Adaptation Method for Online Detection of Bearings Early Fault. *Entropy* 2021, 23, 162. https://doi.org/10.3390/e23020162

Received: 9 January 2021
Accepted: 25 January 2021
Published: 29 January 2021

Publisher's Note: MDPI stays neutral with regard to jurisdictional clai-ms in published maps and institutio-nal affiliations.

Copyright: © 2021 by the authors. Licensee MDPI, Basel, Switzerland. This article is an open access article distributed under the terms and conditions of the Creative Commons Attribution (CC BY) license (https://creativecommons.org/licenses/by/4.0/).

1. Introduction

Early fault detection always plays a key role in the field of bearing prognostics and health management (PHM). In most recent years, the quick development of sensor techniques and artificial intelligence gave rise to a new problem: early fault online detection [1]. Compared with the traditional fault detection and diagnosis problems [2–4], early fault online detection is essentially a problem of anomaly detection with streaming data, that is the monitoring data of the target bearing arrive sequentially, and fault detection is conducted within a sampling interval. This new detection mode can evaluate the change of the working status of bearings in a very short time, avoiding economic losses caused by system halt. Obviously, early fault online detection should not only be sensitive to an early fault, but also be robust enough to avoid false alarms that are usually caused by running-in, lubrication, and so on. Especially, a false alarm can cause unplanned equipment shutdown, so online detection should pay more attention to avoiding false alarms rather than missing alarms. Such characteristics and requirements present a new challenge to the online detection method.

This paper mainly tackles early fault online detection in unsupervised mode, i.e., with no available state information for whole-life degradation data. For online scenarios, a straightforward solution is using the initial part of online data (regarded as the normal state) to construct a one-class classification model. However, a trustful model can usually be built waiting for a long enough time to get sufficient data for model training, especially for a deep neural network. One can certainly accumulate enough whole-life degradation data in an offline environment, e.g., a laboratory. However, the distribution drift between offline data and online data is inevitable due to the change of working conditions. In this scenario, the offline trained model cannot be directly applied to the online task. Therefore, how to transfer fault information (e.g., detection rules) between different working conditions has become a key issue to improve the accuracy and robustness of early fault online detection.

Presently, most traditional fault detection methods heavily rely on the fault features [5] extracted from vibration signals, such as wavelet features [6] and envelop spectrum features [7]. These features are then fed into a classification model such as support vector machine (SVM) [5], naive Bayesian [8], Fisher discrimination analysis [9], artificial neural network [10], and support vector data description (SVDD) [11]. In the past decade, deep learning techniques have been successfully introduced to bearing PHM due to their superior capability of end-to-end feature extraction. As the process of deep feature extraction is self-adaptive with no human intervention, various deep learning techniques have been successfully applied to the fault detection and diagnosis of different rotating machinery [12–15]. However, neither traditional machine learning methods, nor deep learning techniques can effectively solve the problem of distribution drift. Therefore, these methods are not applicable to online detection. According to the authors' literature survey, very few works were found to conduct online anomaly detection. For instance, Lu et al. [16] utilized merely the initial part of online data to build a long short-term memory (LSTM) network and then recognized anomalies by calculating the residual error between real data and the LSTM prediction. Mao et al. [17] utilized semi-supervised SVM and a deep auto-encoder network to sequentially update the classification model for online detection. However, as this method merely used a small amount of normal state data to train the initial model, the extracted deep features are easily biased and then cause a false alarm.

From the discussion above, the most vital challenge to improve the performance of online detection is the effective transfer of fault knowledge between offline and online working conditions. As one of the research hotspots in machine learning, transfer learning aims to improve the predictive performance in one domain (called the target domain) by using the prior information contained in the data of another related, but different domain (called the source domain) [18]. As a kind of transfer learning technique, domain adaptation [19] focuses on the across-domain transfer of domain information. Domain adaptation can be realized well on deep neural networks by adaptively extracting domain-invariant feature representation [20]. Especially in the recent 2–3 years, deep domain adaptation has been applied to fault diagnosis [21–23] and remaining useful life prediction [24,25]. According to our literature survey, there are some preliminary research works [26] in the field of early fault detection. In these works, the role of deep domain adaptation is to learn fault information by leveraging the data from different working conditions. However, there still are some shortcomings: (1) most of these works need labeled data to train a classification model, which is not easy to realize in real-world applications; (2) most of these works mainly focus on anomalous samples rather than the temporal relationship between consecutive samples. As a result, the fault information cannot be sufficiently extracted, which may cause false alarms and reduce detection accuracy.

To solve such shortcomings, this paper proposes a new online detection method of bearing early fault based on deep transfer learning techniques. Specifically, this method contains an offline stage and an online stage. In the offline stage, a new state assessment method is firstly proposed by integrating the Hilbert–Huang transform (HHT) and support vector data description (SVDD) to determine the period of the normal state and the degradation state. The assessment results can provide the corrected data label for further domain adaptation. Furthermore, a new deep dual temporal domain adaptation (DTDA) model is proposed to extract temporal common fault information between different working conditions. In the online stage, the sequentially collected monitoring signals are directly fed into the DTDA model to recognize if a fault occurs. This process does not need a re-training model, since the domain-invariant feature representation has been extracted by the DTDA model. Finally, a set of comparative experiments is conducted on the IEEE PHM Challenge 2012 bearing dataset, and the results demonstrate the effectiveness of the proposed method.

The main contributions of this paper can be summarized as follows:

(1) This paper proposes a new dual temporal domain adaptation model with a dual adaptation strategy. Different from most current deep transfer learning techniques, this model can transfer temporal information of degradation sequences by integrating the time convolutional network (TCN) [27] and the domain adversarial neural network (DANN) [28]. This model can enlarge the temporal characteristics in domain-invariant feature representation and then raise the discrimination between the early fault feature and the normal state feature. According to the authors' best knowledge, there are very few fault detection methods based on transfer learning with temporal information.

(2) This paper presents a new online health indicator (HI) construction method of bearings. This method adopts the temporal common features extracted by the DTDA model and uses principal component analysis (PCA) [29] to get a one-dimensional component. As the extracted common features are representative of the online working condition, the obtained HI can effectively describe the degradation process and provide an intuitive evaluation of online detection results.

This paper is organized as follows. In Section 2, a brief summary about TCN and DANN is provided. In Section 3, the details of the proposed method are elaborated. Section 4 is devoted to showing the experimental results on a widely-used bearing dataset, the IEEE PHM Challenge 2012 dataset, followed by the conclusion of this paper in the last section.

2. Background

2.1. Introduction of the TCN

Rooted in the convolutional neural network (CNN) [30], the TCN has been proven equal to or even better than the recurrent neural network (RNN) [31] in dealing with temporal data [27]. The TCN is mainly composed of three parts: causal convolution, dilated convolution, and residual module, as depicted in Figure 1. A detailed introduction of each part will be given as follows.

Causal convolution is a one-way structure, i.e., the value at time t of the upper layer only depends on the value at time t and before time t of the next layer [27], as shown in Figure 1a. Causal convolution brings the time constraint structure. Dilated convolution is designed to solve the problem that the modeling length with temporal data is restrained by the size of the convolution kernel. In Figure 1b for example, the dilated factor $d = 1$ in the first layer means that every sample is calculated in convolution. If $d = 4$, all four samples are calculated together. As a result, the TCN can obtain a larger receptive field through fewer layers. Here, the function F of dilated convolution with the element s of the sequence X is shown as:

$$F(s) = (X * df)(s) = \sum_{i=0}^{k-1} f(i) \cdot X_{s-d \cdot i} \tag{1}$$

where f is the convolution operation and k is the size of the convolutional kernel.

As shown in Figure 1c, the residual module contains two layers of dilated causal convolution and ReLU mapping. Besides, the TCN runs the dropout after each convolution layer to achieve regularization. The residual module can be expressed as:

$$\hat{z}^{(i)} = \hat{z}^{(i-1)} + F(s) \tag{2}$$

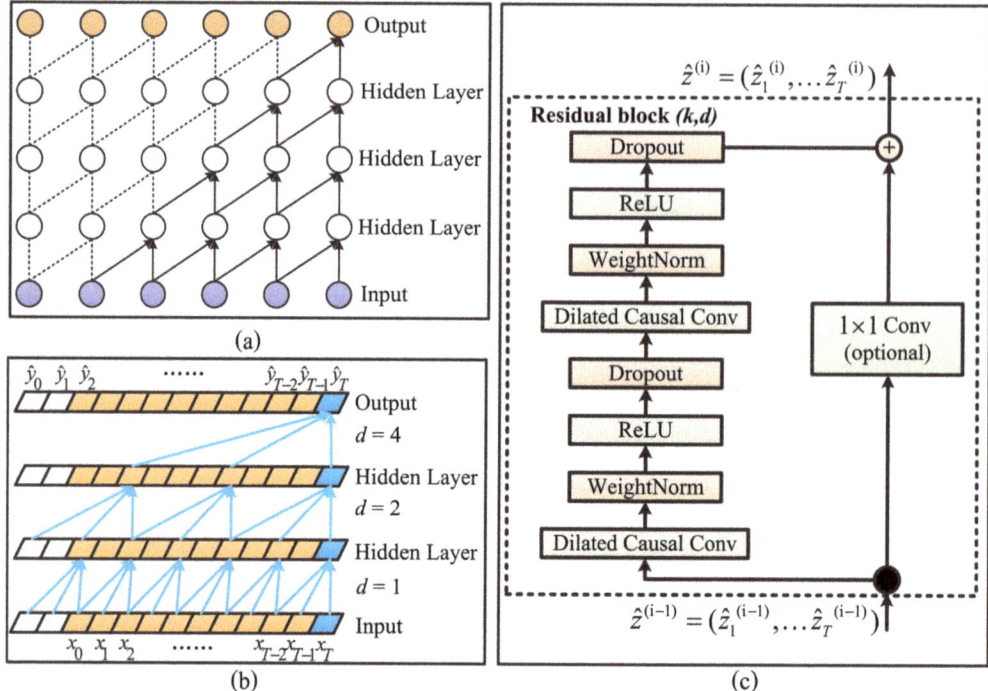

Figure 1. Structure of the time convolutional network (TCN) with (**a**) causal convolution, (**b**) dilated convolution, and (**c**) the residual module.

2.2. Introduction of the DANN

Proposed by Ganin et al. [28], the DANN has become an overwhelming domain adaptation model. The structure of the DANN is shown in Figure 2. In Figure 2, given source domain samples (x_i^s, y_i) and domain samples (x_i, d_i), the green part is a feature extractor $G_f(\cdot; \theta_f)$, the blue part is a source domain classifier $G_y(\cdot; \theta_y)$, and the red part is a domain classifier $G_d(\cdot; \theta_d)$. The adversarial training strategy means that $G_y(\cdot; \theta_y)$ can recognize the data from the source domain using the features extracted by $G_f(\cdot; \theta_f)$ while ensuring that $G_d(\cdot; \theta_d)$ cannot recognize from which domain the data come.

The training process of the DANN mainly concentrates on optimizing $G_y(\cdot; \theta_y)$ and $G_d(\cdot; \theta_d)$. Then, the loss function of $G_y(\cdot; \theta_y)$ is:

$$L_y^i(\theta_f, \theta_y) = L_y(G_y(G_f(x_i^s; \theta_f); \theta_y), y_i) \tag{3}$$

The loss function of $G_d(\cdot; \theta_d)$ is:

$$L_d^i(\theta_f, \theta_d) = L_d(G_d(G_f(x_i; \theta_f); \theta_d), d_i) \tag{4}$$

By adding a gradient reversal layer between $G_y(\cdot; \theta_y)$ and $G_d(\cdot; \theta_d)$, the total loss function of the DANN becomes:

$$\begin{aligned} E(\theta_f, \theta_y, \theta_d) &= \sum_{i=1,2,\ldots N} L_y(G_y(G_f(x_i; \theta_f); \theta_y), y_i) - \lambda \sum_{i=1,2,\ldots N} L_d(G_d(G_f(x_i; \theta_f); \theta_d), d_i) \\ &= \sum_{i=1,2,\ldots N} L_y^i(\theta_f, \theta_y) - \lambda \sum_{i=1,2,\ldots N} L_d^i(\theta_f, \theta_d) \end{aligned} \tag{5}$$

Using a back-propagation optimization algorithm like the stochastic gradient descent (SGD) algorithm, Equation (5) can be minimized to reach equilibrium. Then, a domain-invariant feature representation can be determined. The test data can then be classified using $G_y(G_f(x_i;\theta_f);\theta_y)$.

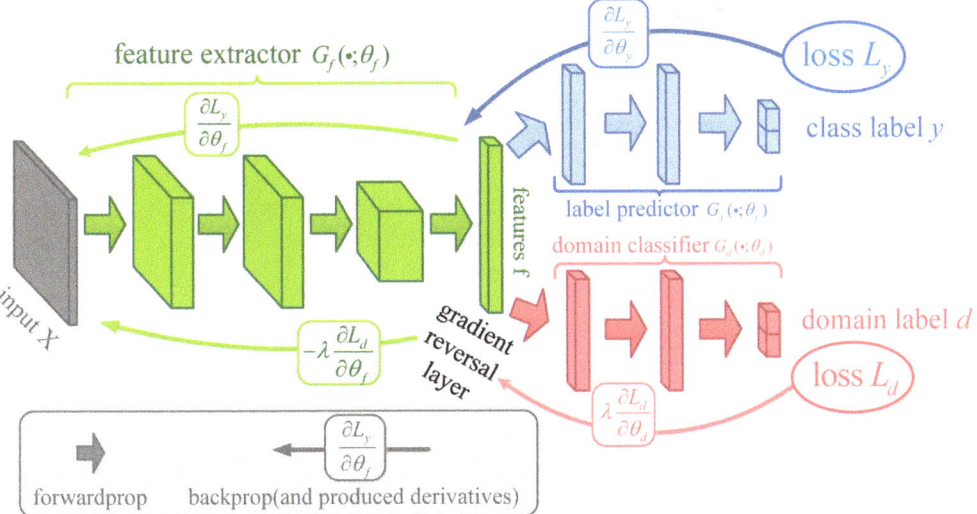

Figure 2. Schematic diagram of the domain adversarial neural network (DANN).

3. Proposed Approach of Early Fault Online Detection

It is worth noting that the proposed method is unsupervised, i.e., all training data are whole-life degradation sequences with no state labels. The training data include a sufficient amount of data from offline working conditions (for example, in a laboratory) and a small amount of data from online working conditions (for example, in a real application). The goal of the proposed method is to recognize the occurrence of an early fault in the online data of a target bearing.

To reach this goal, the proposed method contains an offline stage and an online stage. In the offline stage, it needs to first grab the label information of offline data via state assessment and then extract the domain-invariant feature representation between the offline and online working conditions. Following this idea, a new state assessment method and a novel deep domain adaptation model named DTDA are proposed. In the online stage, the sequentially collected data batch is simply fed into such a feature representation to get the discriminative features for the online tasks and directly get the detection results. The steps stated above are shown in Figure 3.

3.1. State Assessment

The offline data are a whole-life degradation sequence, so we need to determine the label information of the normal state and early fault before conducting domain adaptation. Therefore, an efficient state assessment method is first presented in this section. Given a raw vibration signal sequence of a rolling bearing, the steps of the state assessment are as follows:

(1) Obtain the HHT [32] marginal spectrum data from the raw signal. First, decompose the raw signal $x(t)$ as: $x(t) = \sum_{i=1}^{k} c_i(t) + r_k(t)$, where $c_i(t)$ is the i-th intrinsic mode function (IMF) component and $r_k(t)$ is the residual term. Second, run the Hilbert transform for each IMF component: $H[x(t)] = \frac{1}{\pi} \int_{-\infty}^{+\infty} \frac{x(\tau)}{t-\tau} d\tau$, and get the corresponding analytic signal: $C_i^A(t) = c_i(t) + jc_i^H(t) = a_i(t)e^{j\theta_i(t)}$, where $c_i^H(t) =$

$\frac{1}{\pi}\int_{-\infty}^{+\infty}\frac{c_i(s)}{t-s}ds$, $a_i(t) = \sqrt{c_i^2 + (c_i^H)^2}$, $\theta_i(t) = arctan(c_j^H/c_i)$, with the instantaneous frequency $\omega = \frac{d\theta(t)}{dt}$. Third, calculate the Hilbert spectrum: $H(\omega, t) = \sum_{i=1}^{n} a_i(t)e^{j\theta_i(t)}$, and obtain the marginal spectrum: $H(\omega) = \int H(\omega, t)dt$.

Here, the HHT is regarded as a signal processing method, as the HHT has two merits for analyzing signals: (1) no need to preset the orthogonal basis and (2) the good capability of processing the non-stationary signal. Therefore, here, the HHT is chosen as the signal processing method.

(2) Select the initial 500 samples of the whole-life degradation sequence. Set the HHT marginal spectrum of these samples as the normal state data, and train an SVDD model. Specifically, the optimization target of the SVDD is:

$$\min_{a,R,\xi} R^2 + C \sum_{i=1}^{n} \xi_i \quad (6)$$
$$s.t. \|\phi(x_i - a)\|^2 \leq R^2 + \xi_i, \xi_i \geq 0, \forall i = 1, 2, \ldots n$$

where ξ is the slack variable, R and a are the radius and center of the hyper-sphere, and C is the regularization parameter.

(3) Select the sample sequentially from the beginning and feed the spectrum of the sample into the obtained SVDD. Calculate the distance between this sample and the hyper-sphere center of the SVDD:

$$d = \sqrt{K(x_{test}, x_{test}) - 2\sum_{i=1}^{n}\alpha_i K(x_{test}, x_i) + \sum_{i=1}^{n}\sum_{j=1}^{n}\alpha_i\alpha_j K(x_i, x_j)} \quad (7)$$

where $K(x_i, x_j)$ is the kernel function and α_i is the Lagrange coefficient. If $d \leq R$, the sample x_{test} is recognized as in the normal state, otherwise it is a fault sample. As a result, the boundary between the normal state and the fault state can be determined.

3.2. Proposed DTDA Model

To realize an effective domain adaptation between offline and online working conditions, the DANN is chosen as the baseline algorithm. The DANN adopts the adversarial training strategy and can get a better domain-invariant feature representation even between quite different domains [28]. As training the DANN requires label the information of the source domain, the results of the state assessment presented in Section 3.1 can be used. Different from mature fault data, early fault data generally have a temporal characteristic that reflects the degradation process from the normal state to the fault state. More importantly, the degradation part of the early fault is similar between different bearing sequences [1]. Therefore, the effect of domain adaptation by the DANN can be further enhanced by extracting common temporal information. To extract temporal information well, the TCN is adopted as the feature extractor in the classical DANN, and we propose a new DTDA model.

Specifically, a strategy of dual adaptation is proposed in the DTDA model. This strategy comes from the following observations: (1) Different domains have different requirements on the amount of temporal information, e.g., degradation length. Then, the TCN may perform poorly due it not having a sufficiently large receptive field. (2) The adversarial training strategy used in the DANN may perform unstably when tackling the data with a large distribution difference. Following this analysis, an adaptation layer with the maximum mean discrepancy (MMD) [33] is first added after the TCN's residual blocks. This layer can shrink the distribution difference of temporal features between the source domain and the target domain to some extent. Then, the DANN is run based on such adapted TCN features and can improve the stability of the DANN training as well.

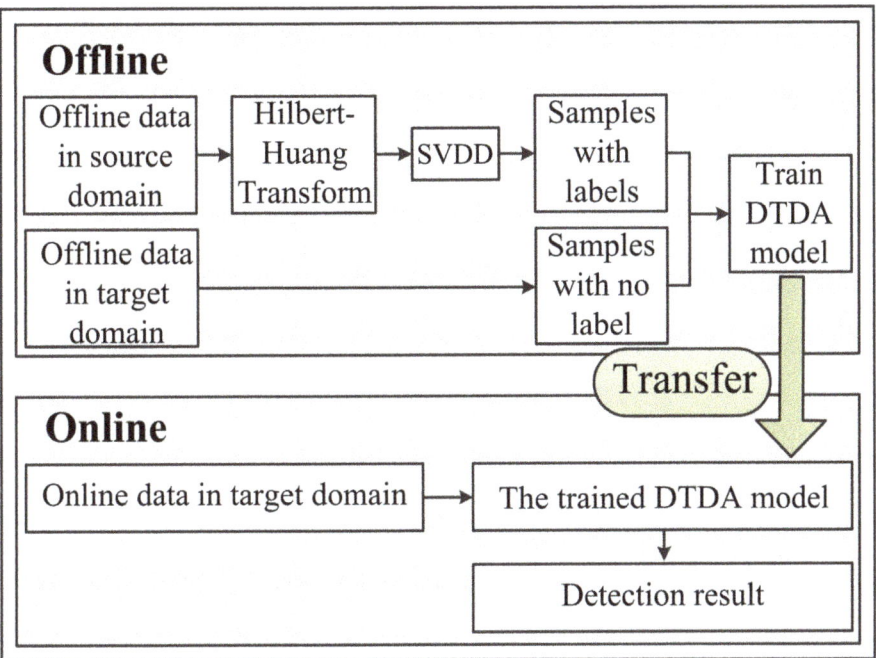

Figure 3. Flowchart of the proposed online detection method of early fault. SVDD, support vector data description; DTDA, dual temporal domain adaptation.

The above idea is shown in Figure 4. Specifically, the orange part represents the source data with labels that are obtained by the state assessment in Section 3.1. The purple part represents the available input data in the target domain. The green part is the feature extractor using the TCN, linked by an MMD adaptation layer. The blue part is the source domain label classifier that aims to recognize the normal state data from the fault data. The pink part is the domain classifier whose task is to discriminate the source domain data and the target domain data. The blue part and pink part are the same as the ones in Figure 2.

The training process of the DTDA model can be summarized as follows:

Step 1. Initialize randomly the weight w and bias b.

Step 2. Combine the source domain data and the target domain data as a whole, and feed them into the TCN to get the output:

$$H_1 = G_f(\sum_{i=1}^{m+n} w_i x_i - b) \tag{8}$$

where m and n are the sample number in the source domain and target domain, respectively, and $G_f(\cdot)$ is the feature extractor of the TCN.

Step 3. Denote by X^s and X^t the source domain feature and target domain feature in H_1, respectively. Then, realize the domain adaptation for X^s and X^t by using an MMD layer. The definition of the MMD is as follows:

$$MMD(X^s, X^t) = \left\| \frac{1}{m}\sum_{i=1}^{m} \phi(x_i^s) - \frac{1}{n}\sum_{j=1}^{n} \phi(x_j^t) \right\|_{\mathcal{H}} \tag{9}$$

where the function $\phi(\cdot)$ indicates a nonlinear mapping to a reproducing kernel Hilbert space (RKHS) and the subscript $\mathcal{H}$ refers to this RKHS.

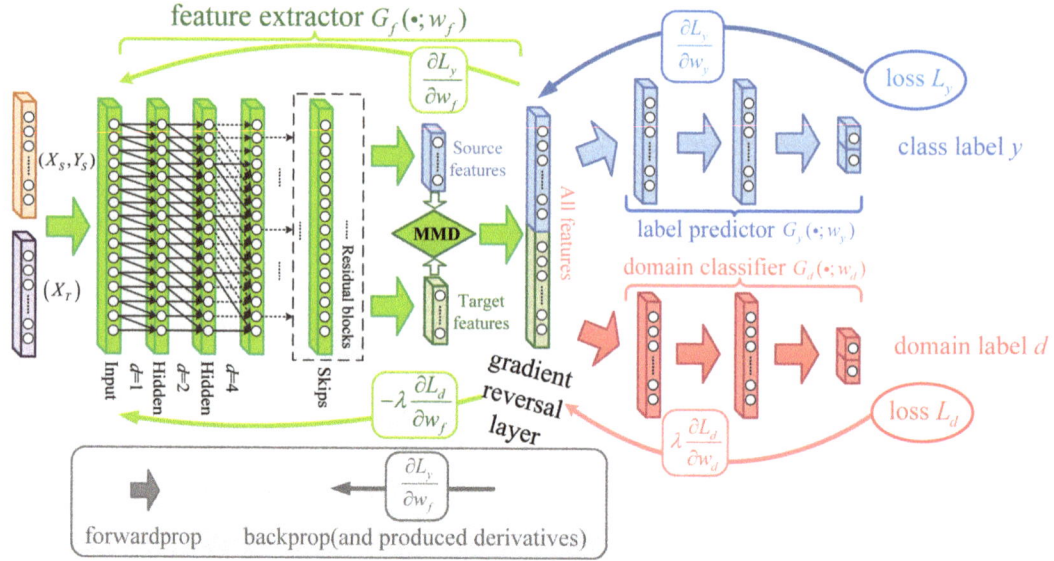

Figure 4. Structure diagram of the proposed DTDA model.

Step 4. Denote by X^s_{MMD} the source domain feature set adapted by the MMD layer. Feed X^s_{MMD} into the source domain label classifier $G_y(\cdot)$ in Figure 4, and get the output $G_y(X^s_{MMD}; w_y)$. The loss function of $G_y(\cdot)$ can be expressed as:

$$L^i_y(w_f, w_y) = L_y(G_y(X^s_{MMD}; w_y), y_i) \qquad (10)$$

where w_f is the model parameter of the feature extractor composed of the TCN and MMD adaptation layer and w_y is the model parameter of the source domain label classifier.

Step 5. Feed the adapted feature set X_{MMD} of the source and target domains into the domain classifier $G_d(\cdot)$, then get the output: $G_d(X_{MMD}; w_d)$. The loss function of $G_d(\cdot)$ can be expressed as:

$$L^i_d(w_f, w_d) = L_d(G_d(X_{MMD}; w_d), d_i) \qquad (11)$$

where w_d is the model parameter of the domain classifier.

Step 6. After combining Equations (9)–(11), the optimization function of the DTDA model is:

$$E(w_f, w_y, w_d) = L^i_y(w_f, w_y) - \lambda L^i_d(w_f, w_d) + \mu MMD(X^s, X^t) \qquad (12)$$

where $\lambda, \mu > 0$ are the regularization parameters, which are used to tune a trade-off between these three quantities during the learning process. Specifically, the larger the value of μ is, the higher the requirement for extracting common features is, and vice versa. Similarly, if the value of λ becomes smaller, the effect of the domain classifier is equivalent to being enhanced, and correspondingly, the samples are more difficult to recognize from the source domain or the target domain. In Section 4, a reverse cross-validation approach, which was adopted in the original DANN model [28], is employed to update the regularization parameters λ and μ. It is worth noting that the minus sign in Equation (12) means gradient reversal for reducing the distribution difference between source domain features and target domain features.

Another thing that needs to be noticed is that the classifier parameters of w_y and w_d are optimized in order to minimize their error on the training set; the feature extractor parameter of w_f is optimized in order to minimize the loss of the source domain label classifier and to maximize the loss of the domain classifier; and the SGD algorithm is employed to minimize Equation (12) to update all three of them.

Step 7. In the training process, if the iteration number reaches a pre-defined number ρ or the difference between two consecutive training errors is less than a pre-defined threshold, the training is terminated; otherwise, go to Step 6.

After dual adaptation, i.e., MMD adaptation and the DANN, the feature distribution of the source domain data and target domain data tends to be consistent. After reaching convergence, the DTDA model can extract the common temporal feature representation of different domain data. In the experiment of this paper, the offline working condition is set as the source domain, and the online working condition is set as the target domain, then such a common feature representation can provide a channel to transfer fault information from the offline data to the online task.

3.3. Online Detection

Once the DTDA is trained, the common temporal feature representation can be obtained. The source domain label classifier $G_y(\cdot)$ can then be used well to recognize the fault in the data of the target domain. Therefore, in the online stage, the sequentially collected data batch is directly fed into $G_y(\cdot)$ to determine if a fault has occurred. This process does no need to re-train the DTDA, while the main computational cost is the linear calculation in $G_y(\cdot)$. As a result, the detection speed is very fast.

3.4. HI Construction

To intuitively evaluate the reliability of the detection results, a new HI construction method is also proposed based on the DTDA model. This method is simple and effective. Specifically, through feeding sequentially the online data batch into $G_y(\cdot)$, not only the detection results, but also the temporal features of the target bearing can be obtained. After detecting all the online data, the features of the whole degradation sequence can be obtained. Then, PCA is run to get the first principal component. After the smoothing operation, the obtained feature sequence is the HI of the target bearing. Since the feature extractor in the DTDA model can extract domain-invariant feature representation with strong discriminative ability, the HI constructed based on such a feature representation can be more sensitive to reflect various state changes of the target bearing. Certainly, the obtained HI can also verify the reliability of the online detection results.

4. Experimental Results

To verify the effectiveness of the proposed method, a set of comparative experiments is run on the IEEE PHM Challenge 2012 bearing dataset [34] in this section. The programming environment was Python 3.6 and MATLAB R2014. The experiments used the Windows operating system (OS) with an i5-7300 processor and 8 G memory.

4.1. Dataset Description

The IEEE PHM Challenge 2012 dataset was collected from PRONOSTIA test platform, as shown in Figure 5, on which an accelerated degradation experiment was conducted to collect run-to-failure data within a few hours. The PRONOSTIA platform is composed of three parts: rotating part, load part, and data collection part. The rotating part has a motor with a power of 250 W. To accelerate degradation, the load part provides a 4000 N load for the rolling bearing. Vibration signals were collected using an accelerometer sensor placed in the horizontal direction. The sampling frequency was 25.6 kHz, while the data were recorded every 10 s. In total, seventeen bearings were selected to collect whole-life degradation data under three working conditions. The specific information of the working conditions is shown in Table 1.

In this experiment, the seven bearings (i.e., Bearing1_1 to Bearing1_7) under the first working condition were selected as the source domain data. Moreover, Bearing 2 and Bearing 3 under the second working condition (i.e., Bearing2_2 and Bearing2_3) were taken as the offline data in the target domain, and we took Bearing 1 (i.e., Bearing2_1) and

Bearing 4 (i.e., Bearing2_4) under the second working condition as the target bearings to be tested in the target domain.

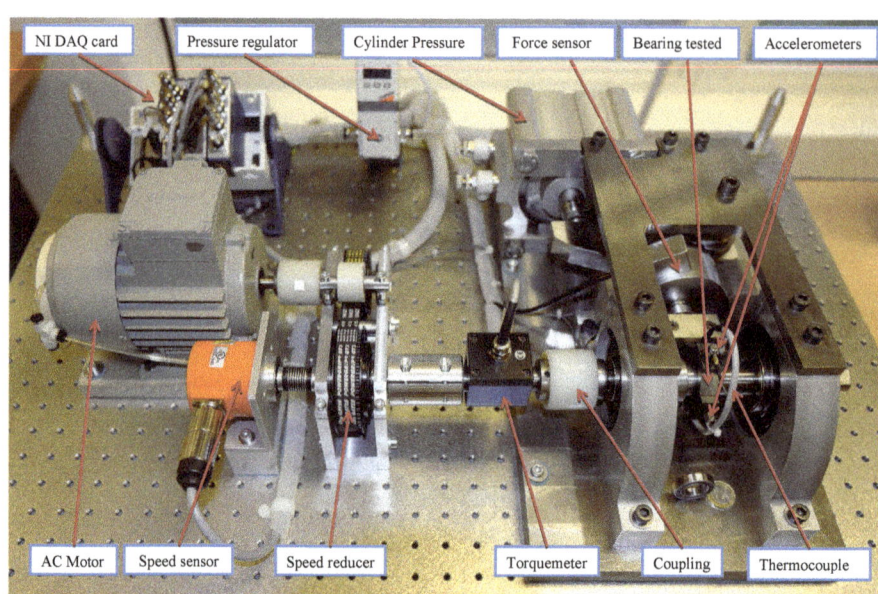

Figure 5. PRONOSTIAtest platform [34].

Table 1. Description of the three working conditions in IEEE Prognostics and Health Management (PHM) Challenge 2012 dataset.

Operating Condition	Rotating Speed (rpm)	Radial Force (N)
First operating condition	1800	4000
Second operating condition	1650	4200
Third operating condition	1500	5000

4.2. Results of State Assessment

In this section, the results of the state assessment are provided. Taking Bearing 1_1 as an example, HHT was first run to get the marginal spectrum data for this bearing, and then, we chose the first 500 samples to train an SVDD model. The Gaussian radial basis function (RBF) kernel was adopted, and the regularization parameter and kernel parameter of SVDD were set to one and 0.001, respectively. After feeding the HHT spectrum data into the trained SVDD model sequentially, the results of the state assessment can be obtained. Table 2 shows the period of the normal state and the fault state of all seven bearings under the first working conditions. These results will be used as the label information for training a DTDA model in the next section.

4.3. Results of Online Detection

In this section, Bearing2_1 and Bearing2_4 under the second working condition are chosen as the target bearings to evaluate the effectiveness of the proposed method. Specifically, these two bearings have quite different degradation trends and noise levels in the normal state. Bearing2_1 has a long period of slow degradation, while Bearing2_4 has no apparent early fault state and quickly evolves to the fast degradation state. Therefore, these two bearings are believed to be representative enough to provide a comprehensive evaluation.

Table 2. State assessment results of the IEEE PHM Challenge 2012 dataset.

Bearing	Normal State Period	Fault State Period
Bearing1_1	[1–1371]	[1372–2803]
Bearing1_2	[1–716]	[717–871]
Bearing1_3	[1–1165]	[1166–2375]
Bearing1_4	[1–931]	[932–1428]
Bearing1_5	[1–2235]	[2236–2463]
Bearing1_6	[1–1587]	[1588–2448]
Bearing1_7	[1–2067]	[2068–2259]

4.3.1. Results of Bearing2_1

First, Figure 6 provides the visualized feature distribution after domain adaptation by DTDA. Here, two bearings (Bearing1_2 and Bearing1_3) in the source domain and two bearings (Bearing2_2 and Bearing2_3) in the target domain are chosen. For comparison, Figure 6 also provides the feature distribution by using the deep autoencoder (DAE) without domain adaptation. Here, PCA is used for visualization. From Figure 6b, before domain adaptation, the feature distribution of the bearings in the source domain (red points and blue points) and the target domain (purple points and green points) vary largely, which indicates that different working conditions have different data distribution characteristics. However, after domain adaptation by the DTDA, the feature distribution of different domains tends to be consistent, as shown in Figure 6a. The results shown in Figure 6 demonstrate that the DTDA model can effectively extract domain-invariant feature representation between different working conditions.

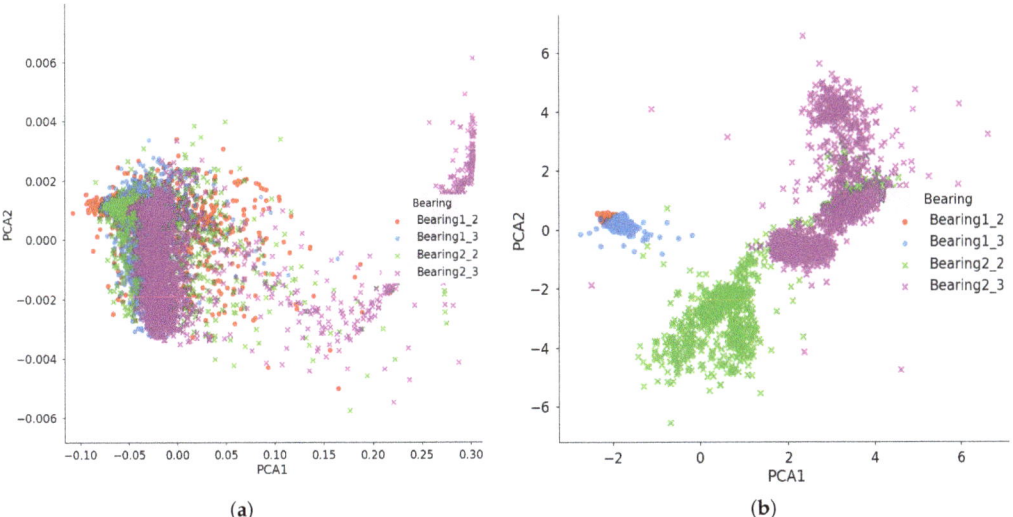

Figure 6. Feature distribution of the four bearings under the first and second working conditions extracted by (a) the DTDA and (b) the deep autoencoder (DAE). Here, PCA is used for visualization.

Second, Figure 7 provides the results of early fault online detection on Bearing2_1. To ensure the results are more reliable, the location of five successive anomalous samples is defined as the occurrence of an early fault. The anomaly before the occurrence location is defined as a false alarm. For straightforward comparison, Figure 7 also reports the HI sequence built by the proposed method in Section 3.4 and the root mean square (RMS) curve. From Figure 7a, an early fault occurs at Sample 162 with only four false alarms. From Figure 7b, the HI sequence has a basically consistent trend with Figure 7a, which proves that the HI sequence can be used to evaluate the reliability of the detection results. As a widely-used indicator to reflect the degradation trend, the RMS curve rises slowly at Sample 180, which lags by nearly 20 samples. This comparison demonstrates that the domain-invariant feature representation extracted by the DTDA model has a better discriminative ability.

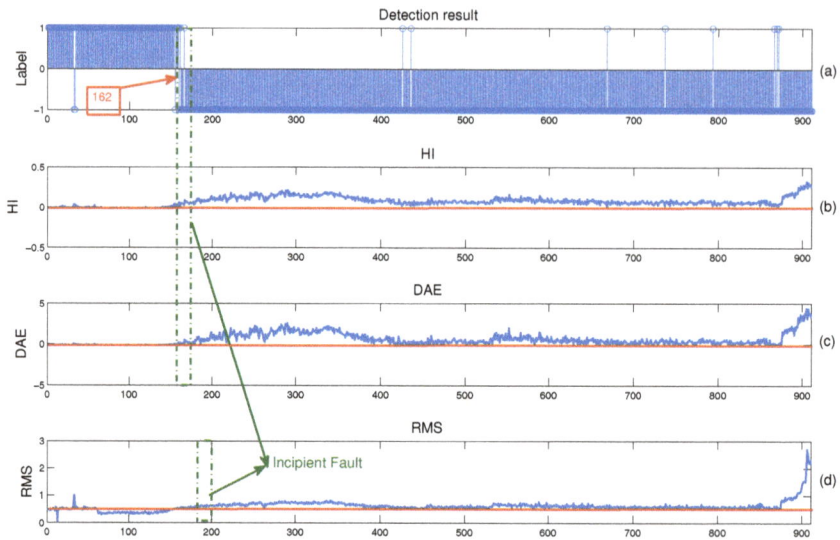

Figure 7. Online anomaly detection results on Bearing2_1 using (**a**) the proposed method, (**b**) the constructed health indicator (HI) by PCA, (**c**) the constructed HI by DAE, and (**d**) the RMS curve. Here, the label "1" in Subfigure (**a**) indicates the normal state; "−1" indicates the fault state.

The effectiveness of the obtained HI sequence is further analyzed. In Section 3.4, PCA is used to shrink the degradation features into a one-dimensional component, which performs as an HI. To test the effectiveness of PCA in HI construction, the DAE is also introduced to build an HI sequence by replacing PCA, as shown in Figure 7c. Specifically, after extracting the domain-invariant feature representation, the features of Bearing2_2 and Bearing2_3 in the target domain can be generated. Then, a DAE model with one-dimensional output is trained using the feature set of these two bearings. Finally, the online features of Bearing2_1 are directly fed into the obtained DAE model to get a one-dimensional output, i.e., the expected HI sequence. It is clear that the two HI sequences built by PCA and DAE are nearly identical in geometric shape, and the location of the early fault is almost the same. This phenomenon indicates that the common features obtained by the DTDA have good representative capability to reflect the degradation trend, while both PCA and the DAE can easily extract a representative component from the features to build the HI. Still, the HI by PCA is a little more sensitive to the early fault. Since training a DAE model generally needs a sufficient amount of data, less samples in the online stage may cause over-fitting. Moreover, the DAE is trained by a gradient descent algorithm, which has more computational cost than PCA. Under comprehensive consideration, PCA is believed to be more suitable for HI construction than the DAE.

In Figure 7d, the RMS curve fluctuates drastically in the initial part, even locating in the normal state. This phenomenon is mainly caused by the irregular vibration of running-in, assembly errors, etc., not by early fault. If the features are not representative (like RMS), there will be many false alarms in the normal state. Quite different from the RMS curve, the HI sequence has almost no irregular fluctuations in the normal state. This phenomenon shows that the DTDA model can extract fault features that are robust to the irregular fluctuations in the normal state. Moreover, the HI sequence has an obvious upward trend after the location of the early fault, while the RMS curve keeps flat for a long period. It is clear that the features extracted by the DTDA are more sensitive to early fault than the RMS feature. As a result, the DTDA model can generate deep features with better discriminative ability, which is helpful to improve the performance of early fault detection.

To further analyze the comparative advantage of the proposed method, the trained DTDA in the offline stage is used to generate the online features of Bearing2_1, as shown in Figure 8. It is worth noting that the visualization label in Figure 8 corresponds to the results in Figure 7a. The features of the two states are almost linearly separable, which indicates that the features extracted by DTDA are discriminative for early fault and very applicable for online detection.

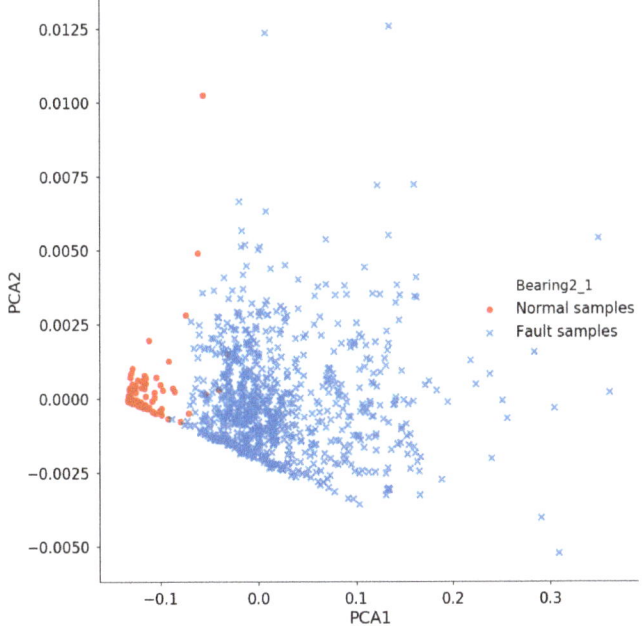

Figure 8. Online features of the target bearing Bearing2_1 in the IEEE PHM Challenge 2012 dataset.

4.3.2. Results of Bearing2_4

First, similar to Figure 7, Figure 9 shows the results of the online detection for Bearing2_4. This bearing falls into the early fault state at Sample 744, with no false alarm. It is also clear that the trend of the HI sequence and the RMS curve completely matches the detection results in Figure 9a, as shown in the dotted frame. Moreover, compared with the RMS curve, the HI sequence does not have an obvious fluctuation, which proves the effectiveness of the proposed method in early fault online detection.

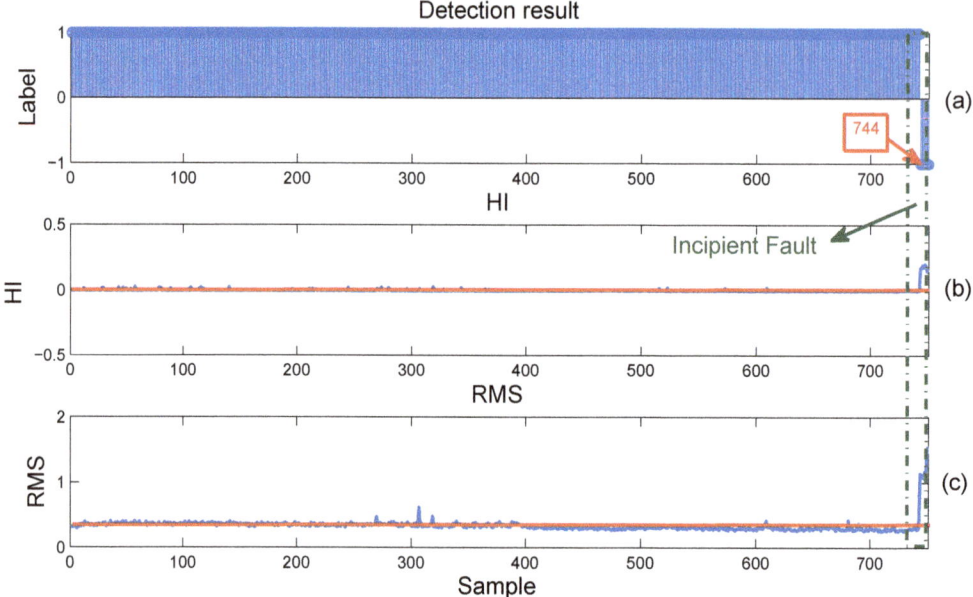

Figure 9. Online anomaly detection results for Bearing2_4 using (**a**) the proposed method, (**b**) the constructed HI, and (**c**) the RMS curve. Here, the label "1" in Subfigure (**a**) indicates the normal state; "−1" indicates the fault state.

Figure 10 further shows the online features of Bearing2_4, which were generated using the domain-invariant feature representation extracted by the DTDA. The separability of these features is even more obvious than the online features of Bearing2_1 shown in Figure 8. This phenomenon is caused by the degradation process of Bearing2_4 in which the bearing transitions directly from the normal state to the fault state. In this scenario, the fault state data are certainly easy to distinguish from the normal state data. Figure 10 proves again that the proposed method can effectively recognize the normal state and the early fault state.

4.4. Comparative Results with State-of-the-Art Methods

In this section, nine state-of-the-art methods of bearing fault detection are introduced for a comprehensive comparison. These nine methods include one typical signal analysis method (Method 1), five anomaly detection methods without transfer learning (Methods 2–6), and three anomaly detection methods with transfer learning (Methods 7–9). For simplicity, the proposed method is named DTDA.

Following [16], two evaluation metrics are employed: (1) the detection location, which is the location (number) of the signal snapshot of the appearing fault; (2) the number of false alarms, which is the number of anomalies before the detection location. The comparative results are reported in Table 3.

From Table 3, the proposed method DTDA obtains the earliest detection location and almost the lowest number of false alarms. Although RD-DTL and SDFM have a lower number of false alarms than the DTDA, the detection location of these two methods is relatively late. It is worth noting that the detection locations of all ten methods on Bearing2_4 are not much different. This is because the bearing evolves quickly from the normal state to the fast degradation state, with a very short period of early fault. Since the data of the fast degradation state are quite different from the normal state data, all methods can detect faults at the location of the state change. However, the number of false alarms produced by different methods on Bearing2_4 is not the same. Some methods like iFOREST and the local outlier factor (LOF) produce too many false alarms. Moreover, bandwidth empirical mode decomposition-adaptive multi-scale morphological analysis (BEMD-AMMA) has no false alarm since it utilizes signal analysis to conduct fault detection by observing the fault frequency.

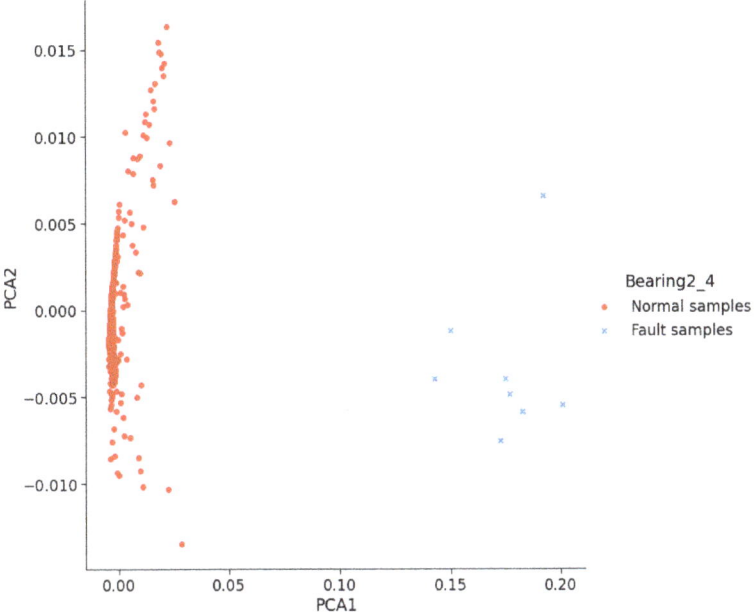

Figure 10. Online features of the target Bearing2_4. Here, PCA is used for visualization.

Table 3. Comparative results of the proposed method with nine state-of-the-art methods. Earlier detection location and lower number of false alarms indicate better. BEMD-AMMA, bandwidth empirical mode decomposition-adaptive multi-scale morphological analysis; LOF, local outlier factor; TCA, transfer component analysis.

Type of Methods	Methods	Bearing2_1		Bearing2_4	
		Detection Location	Number of False Alarms	Detection Location	Number of False Alarms
Signal analysis	1. BEMD-AMMA [35]	185	–	748	–
Anomaly detection without transfer learning	2. LOF [36]	610	40	746	137
	3. iFOREST [37]	307	35	745	152
	4. SRD [38]	870	12	745	13
Online anomaly detection without transfer learning	5. SDFM [1]	175	0	744	0
	6. S4VM-SODRMB [17]	763	10	744	11
Anomaly detection with transfer learning	7. TCA+SVDD	615	21	744	46
Online anomaly detection with transfer learning	8. RD-DTL [39]	175	0	744	0
	9. OD-DTL [26]	169	5	744	3
	10. **DTDA**	**162**	**3**	**744**	**0**

Here, a detailed analysis of the comparative results is listed as follows:

(1) Comparison with BEMD-AMMA:
BEMD-AMMA can be viewed as a state-of-the-art signal analysis-based method for bearing early fault detection. This method first uses the bandwidth empirical mode decomposition (BEMD) to reconstruct the raw vibration signal and then utilizes an

adaptive multi-scale morphological analysis (AMMA) algorithm to demodulate the reconstructed signal to obtain time-domain signals. Finally, a fault can be determined if the fault characteristic frequency can be observed. To calculate the fault characteristic frequency, this method has to know the various parameters of the target bearing and working condition in advance. Obviously, this limitation is too strict to achieve in the online scenario. Moreover, a fault only evolves to a certain degree, and the corresponding characteristic frequency can then appear. Therefore, the detection location will be delayed. In contrast, benefiting from the sensitivity of the early fault features extracted by the DTDA model, the proposed method can detect fault occurrence at an earlier location.

(2) Comparison with LOF:

The LOF is a typical anomaly detection algorithm running on sample density. In this experiment, the first 100 samples at the starting online stage were chosen to calculate the LOF value, and then, the largest value is selected as the alarm threshold. The parameter K in the LOF was set to 10. From Table 3, the detection location of the LOF is later than that of the proposed method, while the number of false alarms is much larger. This is because the normal state of a bearing may have unexpected irregular fluctuations. Moreover, when using normal state data to train the LOF, the threshold value will be relatively larger, resulting in a late detection location.

(3) Comparison with iFOREST:

iFOREST is also a typical anomaly detection algorithm adopting a random segmentation strategy. iFOREST segments all samples into various isolated outliers, and the ones with a shorter path are viewed as anomalies. In this experiment, the number of trees was set to 100. From Table 3, the detection location of iFOREST is much delayed for Bearing2_1, and too many false alarms appear. This is because with online samples arriving sequentially, the segmentation number continues to increase. Consequently, the detection performance is not stable.

(4) Comparison with SRD:

SRD is a state-of-the-art early fault detection algorithm based on probability density estimation. In the offline stage, this method directly uses the original signals of multiple working conditions to establish a global model and determine the alarm limit. The online signals are fed into this model to get the detection results. However, this method does not take into account the difference between the bearing data under different working conditions. The setting of the alarm limit is also too subjective. Therefore, the detection location was very much delayed for Bearing2_1. In contrast, the proposed method does not need to manually set a threshold for detection, and the DTDA model can effectively transfer fault information between different working conditions to improve the detection performance.

(5) Comparison with SDFM:

SDFM is a state-of-the-art online detection method for bearing early fault. This method employs a sliding window to determine the location of early fault occurrence by means of DAE features. The DAE network structure was set to [800, 512, 10], and the size of the sliding window was set to 100. Benefiting from the sliding window, this method had no false alarm, but the detection location was a bit delayed. The main reason is that this method is not a transfer learning method and heavily relies on the amount of offline training data. If the amount of training data is insufficient, the early fault information could not be extracted completely. In contrast, the proposed method can borrow data from offline working conditions to supplement early fault information for online detection.

(6) Comparison with S4VM-SODRMB:

This method is also a state-of-the-art online detection method for bearing early fault. This method only needs online data to update model training in an unsupervised learning architecture. In this experiment, the first 100 samples of online data were accumulated to extract the DAE features and then train an initial SVM model. The sequentially collected data batch was used to update the SVM model successively. The radius-margin upper bound of leave-one-out error was then utilized to calculate an index for online detection. In Table 3, the detection location of this method was much delayed for Bearing2_1. The reason is that only a small amount of online data was used to train an initial model. Once the data contain irregular fluctuation or colored noise, the initial model will be biased, and the detection results will deteriorate. In contrast, the proposed method can utilize offline data to facilitate online detection. The transfer learning technique in the DTDA model guarantees the effective use of early fault information from offline data.

(7) Comparison with TCA + SVDD:

Transfer component analysis (TCA) is a widely used transfer learning algorithm by minimizing the MMD distance between different domains. In this experiment, TCA was first run to conduct domain adaptation between the available data from offline and online working conditions. Then, the common features were used to train an SVDD model by using the available training data from the online working condition. This SVDD model was used to recognize anomalies in the online stage. In this experiment, the regularization parameter and kernel parameter were set to 10 and one, respectively. From Table 3, it is clear that the detection location of TCA+SVDD was much delayed for Bearing2_1, and the number of false alarms became larger. The reason is that TCA conducts domain adaptation with a shallow model, while such domain adaptation is in a single mode. In contrast, the proposed method not only conducts dual domain adaptation, but also extracts common temporal information of early fault. Therefore, the proposed method can provide more representative features for online detection.

(8) Comparison with RD-DTL and OD-DTL:

RD-DTL and OD-DTL are both the newest state-of-the-art early fault online detection methods based on deep transfer learning. RD-DTL first uses a robust auto-encoder to determine the periods of different degradation states and then constructs an MMD-based DAE network to extract common features for the normal state, followed by an SVM model for recognition. This method focuses on the robustness of the online detection model. OD-DTL utilizes a pre-trained VGG-16 network on the ImageNet dataset to fine-tune a deep neural network for bearing online detection. This method only conducts model-level domain adaptation, not considering the feature transfer. Therefore, the performance of fault detection is limited. From Table 3, these two methods get similar results as the proposed method. RD-DTL even gets zero false alarms. However, the proposed method gets an earlier detection location than them. The main reason is that the proposed method conducts dual domain adaptation with temporal information. Therefore, the online features by DTDA are more sensitive to early fault.

In summary, the proposed DTDA model can achieve dual domain adaptation at the feature level, which can facilitate the transfer of fault information between different working conditions. Moreover, the DTDA utilizes the TCN as the feature extractor to extract the temporal information of the degradation process, which can improve the representative ability of the online features for early fault. Therefore, the proposed method is more applicable to the online detection of bearing early fault.

Another problem of online detection is computational time. The proposed method needs to train the DTDA model in the offline stage and then directly inputs the sequentially collected online data into the model to recognize the fault occurrence. The dual domain adaptation by the DTDA provides a domain-invariant feature representation with a better discriminative ability for the online task. The offline model training is computationally expensive, since the adversarial training of the DTDA is an iteration process. However, no additional training time is required in the online stage. The classification on an online sample by the trained DTDA model is almost a linear operation, so the time for recognizing an online sample is very short. For this reason, the corresponding time data are not provided in Table 3.

5. Conclusions

Online detection of bearing early fault is an application-oriented fault detection method with significant practical meaning. This paper proposes a new online detection method of early fault based on deep dual temporal domain adaptation. This method adopts deep domain adaptation with temporal information to extract domain-invariant feature representation with stronger discriminative ability. Employing this representation as the channel of information transfer, the proposed method can improve the detection robustness and accuracy in the online scenario with fewer false alarms as well. This method can directly tackle whole-life degradation data, with no need to manually mark fault data in advance. Therefore, this method is more applicable for the online detection of early fault, and the idea of this paper can be widely expanded for different objects.

In the next work, an attention mechanism will be introduced into domain adversarial training to improve the effect of domain adaptation for time series data. Besides, this paper focuses on the anomaly detection problem for a bearing across different working conditions. How to achieve online transfer learning across different machinery and extract common features from multiple sources is an interesting problem.

Author Contributions: Conceptualization, W.M. and B.S.; methodology, W.M.; software, L.W.; validation, B.S. and L.W.; formal analysis, W.M.; investigation, W.M.; resources, B.S.; data curation, W.M.; writing, original draft preparation, B.S.; writing, review and editing, B.S.; visualization, B.S.; supervision, L.W.; funding acquisition, W.M. All authors have read and agreed to the published version of the manuscript.

Funding: This research was funded by the National Natural Science Foundation of China Grant Number U1704158, Henan Province Technologies R & D project of China Grant Number 212102210103 and 202102210361, DOE Key Scientific Research Project of Henan Province Grant Number 20A520039, and the funding scheme of University Young Core Instructor in Henan Province Grant Number 2019GGJS279.

Institutional Review Board Statement: Not applicable.

Informed Consent Statement: Not applicable.

Data Availability Statement: The data presented in this study are openly available in PCoE Datasets at https://ti.arc.nasa.gov/tech/dash/groups/pcoe/prognostic-data-repository/, reference number [34].

Conflicts of Interest: The authors declare no conflict of interest.

References

1. Mao, W.; Chen, J.; Liang, X.; Zhang, X. A new online detection approach for rolling bearing incipient fault via self-adaptive deep feature matching. *IEEE Trans. Instrum. Meas.* **2019**, *69*, 443–456. [CrossRef]
2. Tian, J.; Morillo, C.; Azarian, M.H.; Pecht, M. Motor bearing fault detection using spectral kurtosis-based feature extraction coupled with K-nearest neighbor distance analysis. *IEEE Trans. Ind. Electron.* **2015**, *63*, 1793–1803. [CrossRef]
3. Yang, Y.; Yu, D.; Cheng, J. A fault diagnosis approach for roller bearing based on IMF envelope spectrum and SVM. *Measurement* **2007**, *40*, 943–950. [CrossRef]
4. Guo, X.; Chen, L.; Shen, C. Hierarchical adaptive deep convolution neural network and its application to bearing fault diagnosis. *Measurement* **2016**, *93*, 490–502. [CrossRef]
5. Liu, R.; Yang, B.; Zhang, X.; Wang, S.; Chen, X. Time-frequency atoms-driven support vector machine method for bearings incipient fault diagnosis. *Mech. Syst. Signal Process.* **2016**, *75*, 345–370.
6. Sun, Q.; Tang, Y. Singularity analysis using continuous wavelet transform for bearing fault diagnosis. *Mech. Syst. Signal Process.* **2002**, *16*, 1025–1041. [CrossRef]
7. Sun, W.; An Yang, G.; Chen, Q.; Palazoglu, A.; Feng, K. Fault diagnosis of rolling bearing based on wavelet transform and envelope spectrum correlation. *J. Vib. Control* **2013**, *19*, 924–941.
8. Dhamande, L.S.; Chaudhari, M.B. Compound gear-bearing fault feature extraction using statistical features based on time-frequency method. *Measurement* **2018**, *125*, 63–77. [CrossRef]

9. Li, F.; Wang, J.X.; Chyu, M.K.; Tang, B.P. Weak fault diagnosis of rotating machinery based on feature reduction with Supervised Orthogonal Local Fisher Discriminant Analysis. *Neurocomputing* **2015**, *168*, 505–519. [CrossRef]
10. Amar, M.; Gondal, I.; Wilson, C. Vibration spectrum imaging: A novel bearing fault classification approach. *IEEE Trans. Ind. Electron.* **2015**, *62*, 494–502.
11. Tax, D.M.; Duin, R.P. Support vector data description. *Mach. Learn.* **2004**, *54*, 45–66. [CrossRef]
12. Luo, B.; Wang, H.; Liu, H.; Li, B.; Peng, F. Early fault detection of machine tools based on deep learning and dynamic identification. *IEEE Trans. Ind. Electron.* **2018**, *66*, 509–518. [CrossRef]
13. Shao, H.; Jiang, H.; Li, X.; Liang, T. Rolling bearing fault detection using continuous deep belief network with locally linear embedding. *Comput. Ind.* **2018**, *96*, 27–39. [CrossRef]
14. He, M.; He, D. Deep learning based approach for bearing fault diagnosis. *IEEE Trans. Ind. Appl.* **2017**, *53*, 3057–3065. [CrossRef]
15. Sun, J.; Yan, C.; Wen, J. Intelligent bearing fault diagnosis method combining compressed data acquisition and deep learning. *IEEE Trans. Instrum. Meas.* **2017**, *67*, 185–195. [CrossRef]
16. Lu, W.; Li, Y.; Cheng, Y.; Meng, D.; Liang, B. Early Fault Detection Approach With Deep Architectures. *IEEE Trans. Instrum. Meas.* **2018**, *67*, 1679–1689. [CrossRef]
17. Mao, W.T.; Tian, S.Y.; Fan, J.J.; Liang, X.H.; Safian, A. Online detection of bearing incipient fault with semi-supervised architecture and deep feature representation. *J. Manuf. Syst.* **2020**, *55*, 179–198. [CrossRef]
18. Pan, S.J.; Yang, Q. A survey on transfer learning. *IEEE Trans. Knowl. Data Eng.* **2009**, *22*, 1345–1359. [CrossRef]
19. Pan, S.J.; Tsang, I.W.; Kwok, J.T.; Yang, Q. Domain adaptation via transfer component analysis. *IEEE Trans. Neural Netw.* **2010**, *22*, 199–210. [CrossRef]
20. Tan, C.; Sun, F.; Kong, T.; Zhang, W.; Yang, C.; Liu, C. A survey on deep transfer learning. In *International Conference on Artificial Neural Networks*; Springer: Cham, Switzerland, 2018; pp. 270–279.
21. Tan, Y.; Guo, L.; Gao, H.; Zhang, L. Deep Coupled Joint Distribution Adaptation Network: A Method for Intelligent Fault Diagnosis Between Artificial and Real Damages. *IEEE Trans. Instrum. Meas.* **2020**, *70*, 1–12. [CrossRef]
22. Guo, L.; Lei, Y.; Xing, S.; Yan, T.; Li, N. Deep convolutional transfer learning network: A new method for intelligent fault diagnosis of machines with unlabeled data. *IEEE Trans. Ind. Electron.* **2018**, *66*, 7316–7325. [CrossRef]
23. Wu, Z.; Jiang, H.; Zhao, K.; Li, X. An adaptive deep transfer learning method for bearing fault diagnosis. *Measurement* **2020**, *151*, 107227. [CrossRef]
24. Mao, W.; He, J.; Zuo, M. Predicting remaining useful life of rolling bearings based on deep feature representation and transfer learning. *IEEE Trans. Ind. Electron.* **2020**, *64*, 1594–1608. [CrossRef]
25. Zhang, A.; Wang, H.; Li, S.; Cui, Y.; Liu, Z.; Yang, G.; Hu, J. Transfer learning with deep recurrent neural networks for remaining useful life estimation. *Appl. Sci.* **2018**, *8*, 2416.
26. Mao, W.; Ding, L.; Tian, S.; Liang, X. Online detection for bearing incipient fault based on deep transfer learning. *Measurement* **2019**, *152*, 107278. [CrossRef]
27. Bai, S.; Kolter, J.Z.; Koltun, V. An empirical evaluation of generic convolutional and recurrent networks for sequence modeling. *arXiv* **2018**, arXiv:1803.01271.
28. Ganin, Y.; Ustinova, E.; Ajakan, H.; Germain, P.; Larochelle, H.; Laviolette, F.; March, M.; Lempitsky, V. Domain-adversarial training of neural networks. *J. Mach. Learn. Res.* **2016**, *17*, 1–35.
29. Wold, S.; Esbensen, K.; Geladi, P. Principal component analysis. *Chemom. Intell. Lab. Syst.* **1987**, *2*, 37–52. [CrossRef]
30. Krizhevsky, A.; Sutskever, I.; Hinton, G.E. Imagenet classification with deep convolutional neural networks. *Commun. ACM* **2017**, *60*, 84–90. [CrossRef]
31. Schuster, M.; Paliwal, K.K. Bidirectional recurrent neural networks. *IEEE Trans. Signal Process.* **1997**, *45*, 2673–2681.
32. Yan, H.; Ding, Y.; Li, P.; Wang, Q.; Xu, Y.; Zuo, W. Mind the class weight bias: Weighted maximum mean discrepancy for unsupervised domain adaptation. In Proceedings of the IEEE Conference on Computer Vision and Pattern Recognition, Honolulu, HI, USA, 21–26 July 2017; pp. 2272–2281.
33. Borgwardt, K.M.; Gretton, A.; Rasch, M.J.; Kriegel, H.P.; Schölkopf, B.; Smola, A.J. Integrating structured biological data by kernel maximum mean discrepancy. *Bioinformatics* **2006**, *22*, e49–e57. [CrossRef] [PubMed]
34. Nectoux, P.; Gouriveau, R.; Medjaher, K.; Ramasso, E.; Chebel-Morello, B.; Zerhouni, N.; Varnier, C. PRONOSTIA: An Experimental Platform for Bearings Accelerated Degradation Tests. In Proceedings of the IEEE International Conference on Prognostics and Health Management, PHM'12, Denver, CO, USA, 18–21 June 2012; pp. 1–8.
35. Li, Y.; Xu, M.; Liang, X.; Huang, W. Application of bandwidth EMD and adaptive multiscale morphology analysis for incipient fault diagnosis of rolling bearings. *IEEE Trans. Ind. Electron.* **2017**, *64*, 6506–6517. [CrossRef]
36. Kriegel, H.P.; Kröger, P.; Schubert, E.; Zimek, A. LoOP: Local outlier probabilities. In Proceedings of the 18th ACM conference on Information and Knowledge Management, Hong Kong, China, 2–6 November 2009; pp. 1649–1652.
37. Liu, F.T.; Ting, K.M.; Zhou, Z.H. Isolation forest. In Proceedings of the 2008 Eighth IEEE International Conference on Data Mining, Pisa, Italy, 15–19 December 2008; pp. 413–422.
38. Guo, J.; Liu, S.; Li, Y. Fault detection of multi-mode processes employing sparse residual distance. *Acta Autom. Sin.* **2019**, *45*, 617–625.
39. Mao, W.; Zhang, D.; Tian, S.; Tang, J. Robust Detection of Bearing Early Fault Based on Deep Transfer Learning. *Electronics* **2020**, *9*, 323. [CrossRef]

MDPI
St. Alban-Anlage 66
4052 Basel
Switzerland
Tel. +41 61 683 77 34
Fax +41 61 302 89 18
www.mdpi.com

Entropy Editorial Office
E-mail: entropy@mdpi.com
www.mdpi.com/journal/entropy

www.ingramcontent.com/pod-product-compliance
Lightning Source LLC
LaVergne TN
LVHW070154120526
838202LV00013BA/1093